Arthur E. Kennelly

Electric Street Railways

Arthur E. Kennelly

Electric Street Railways

ISBN/EAN: 9783744678841

Printed in Europe, USA, Canada, Australia, Japan

Cover: Foto ©berggeist007 / pixelio.de

More available books at **www.hansebooks.com**

ELEMENTARY ELECTRO-TECHNICAL SERIES

ELECTRIC STREET RAILWAYS

BY

EDWIN J. HOUSTON, Ph. D.

AND

A. E. KENNELLY, Sc. D.

NEW YORK

THE W. J. JOHNSTON COMPANY

253 BROADWAY

1896

clear conception of its method of operation. The authors have prepared this little volume of the *Electro-Technical Series* in the belief that these difficulties are apparent rather than real—that it is quite possible for the general public to obtain a fairly intimate knowledge of the leading principles of electric traction without any previous knowledge of electrotechnics.

It is a matter of necessity rather than choice, at the close of this nineteenth century, when electric traction has become so nearly universal, that a knowledge of the main principles concerned should be generally accessible, without special training, and more especially is this desirable on the part of those, now an exceedingly extensive class, who are connected in some way or other with such enterprises.

The authors present this book to the general public hoping that it will meet the need above referred to.

CONTENTS.

ELECTRIC STREET RAIL-
WAYS.

CHAPTER I.

INTRODUCTION.

The introduction of the electric street
railway naturally caused much wonder-
ment. There seemed at the first some-
thing weird in the possibility of propelling a heavily loaded vehicle, from place to
place, without any apparent motive power,
and, even at the present time, there remains
no little astonishment in the mind of the
casual observer as to how the electric
agency can silently, yet surely, find its way
from a power house, in some remote corner
of a city, through an intricate maze of

streets and turnings, and propel each car as though the latter were under the guidance of a familiar spirit. The wonder grows, when it is pointed out that the electric current not only has to find its way from the power house over the trolley wires to the cars, wherever these may be, but has also to return to the power house through the track and ground.

It is, unfortunately, too true that the real nature of electricity remains unknown, even in this electric age. For this reason, there has, perhaps, existed, in the minds of the public, too marked an unwillingness to attempt even to form ideas as to the laws which control electric operations. But it should not be forgotten, that although our knowledge of the exact nature of electricity is imperfect, yet our knowledge of the manner in which it operates, that is of the laws which control it, is surprisingly definite. Indeed, so far as the laws which govern the flow of electric currents through conducting paths or circuits are concerned, our knowledge is even more definite than of the laws which control the flow of water or gas through pipes. In fact, as we shall subsequently see, a remarkable analogy exists between the laws which govern the flow of gross matter in the fluid state, that is as liquids or gases, and the laws which govern the flow of electricity.

It may be well, therefore, before proceeding further with the general discussion of electric street railways, to outline briefly the points of similarity between the flow of liquids and the flow of electricity.

Perhaps no better illustration could be given, concerning some of the laws of liquid flow, than that taken from the distribution of water through the mains and pipes of a large city. Here, as is well known, a supply of water is provided in a reservoir, at a high level or pressure. Pipes or mains connecting with this reservoir extend beneath the streets to all portions of the city that are to be supplied with water. No difficulty will be experienced in understanding how, if no obstruction exists in the pipes, the water will flow through them from the reservoir and escape through any outlet at a lower level.

Let us now examine the network of pipes connected with a reservoir in a system of municipal water distribution. It is evident that the object of such a system is to supply the houses or other buildings either with water, or with the power the water is capable of exerting. For this purpose two sets of pipes are provided; viz.,

(1) Those connected immediately with the reservoir and intended to carry the water.

(2) Those connected to the consumers' waste pipes.

The latter are connected intermediately with the sewer system, and ultimately with the lake, river, or ocean into which such sewer system discharges.

Between the reservoir and the river, it is evident that the flow of water through the pipes is due to gravity, the water finding its way through the pipes in obedience to the law of liquid levels. After the river has been reached, and the water is ultimately discharged into the ocean, thus reaching its lowest level, some means must be provided for causing the water to rise against the force of gravity and fill the reservoir afresh. This energy is received from the sun during the evaporation of the water when it passes into vapor and rises into the atmosphere. On the loss of the heat so received the water again falls under the influence of gravity, and returns to the reservoir.

An analogy between the preceding system of water distribution through the pipes of a city, and a system of electric distribution through the trolley wires, is evident. Here, as we shall more fully see in a subsequent chapter, an actual differ- ence of electric level exists, whereby an electric source, or generator, at the power house, causes the electricity to flow through all the conducting outgoing trol- ley wires from the higher electric level of the generator to the cars. In passing through the cars, it may light and heat them as well as drive their motors. On leaving the cars it flows through the ground back again to the generators in the power house. In this latter part of its circuit or path, an analogy is to be found between the discharge of the water to the lower level of the ocean, prior to its passage back again to the higher level of the reservoir.

Although we have thus traced the analogy between liquid flow and electric flow, and have shown that the same general laws apply to each, yet it must be remembered that this is an analogy only; and that electricity is not believed to be a material fluid. The analogy is, however, useful, and will aid the student in forming practical conceptions of the electric circuit.

CHAPTER II.

THE broad idea of propelling vehicles by means of the electric current appears to have suggested itself to the minds of inventors at an early date. As long ago as 1835, Thomas Davenport, of Vermont, constructed a working model of a car propelled by an electric motor of his own invention. In 1838, Robert Davidson, of Scotland, also produced an electrically propelled car. Both of these early cars derived their propelling current from voltaic batteries carried on the car.

The idea of taking the electric current required for the propulsion of the car from

8

conductors laid alongside the track was not conceived until a somewhat later date; namely, in 1840, when Henry Pinkus obtained letters patent in Great Britain for a method of propelling carriages either on railroads or on ordinary highways. This patent discloses among other things, the broad idea of taking electric current from conductors, in contradistinction to employing batteries on the car.

Space will not permit us to enter in detail on this portion of the early history of the electric railway. It will suffice to say, that in 1851, Professor Page of the Smithsonian Institution devised an electric locomotive which he ran on a track at the rate of nineteen miles an hour. This locomotive, like those of Davenport and Davidson, carried the voltaic battery required for its propulsion. About the same time Professor Moses G. Farmer also devised an electrically propelled car.

All these early discoveries belong to the type of ideas that are born too early to come to fruition. Practically the only electric source that was known at this date was the voltaic battery, which is incapable of commercially producing the powerful electric currents required for the propulsion of street cars. It was not until the dynamo-electric machine was perfected that electric car propulsion became commercially practicable.

The advent of the dynamo-electric generator, therefore, marked the second era in the history of electric railway development. The low cost at which this electric source can furnish powerful currents attracted the attention of inventors, who long before had recognized the part electricity was destined to play in electric locomotion. Consequently, this era of the history of the electric railway contains many inventions.

It is not our intention to enter into any discussion as to the claims of the various inventors to priority in any of the more salient features of the art of electric traction. We will content ourselves with a brief account only of some of the work accomplished at this time.

One of the pioneers at the beginning of this era in the history of the electric railway was George Green, who devised a road on a plan similar to that of Farmer, but containing many marked improvements. Green, who was poor, experienced difficulty in getting his patent interests

attended to.   Being placed in interferences
with other applicants, a patent was not
issued to him until the last month of 1891,
although applied for as early as 1879.

Passing by a number of inventors who
devised electric locomotives of various
types, we come to the electric railway of
Siemens and Halske, which was put into
actual operation at the Industrial Exhibi-
tion of Berlin in 1879.   As in all electric
railways belonging to this era, the motive
power was derived from dynamos located
at a central station.   The current was
delivered to the motor by means of a slid-
ing contact under the locomotive, rubbing
against a rail placed midway between the
two track rails.

Very little was done in electric railways
in the United States, prior to 1883.   It is

true that in 1880 some work was undertaken by Edison which resulted in the erection of an experimental track, and that prior to this date; namely, in May, 1879, Stephen D. Field had done some experimental work which he protected in the United States Patent office by a caveat.

In the meantime inventors in other countries had by no means been idle. The honor of establishing the first commercial electric street railway appears to belong to Germany, where the Lichtenfeld line was put in operation in 1881. Another road was opened at Portrush, in the north of Ireland, in 1883, the dynamos being in this case driven by water power.

Among early railways operated in the United States was one constructed and put in operation by Vanderpoele, at the

Chicago State Fair during two months in 1884. A short line, located on one of the piers on Coney Island, N. Y., was operated during the summer season of 1884. The year 1884 also saw the first public electric street railway in operation at Providence, R. I., and the first practical trolley road was that in the suburbs of Kansas City, Mo., in the same year.

The advantages possessed by electric traction over ordinary methods, such for example, as horse cars, are so great, that while in 1884 the first electric road was installed in the United States, there were, in

1889,	50 roads with	100	miles of track.
1890,	200 "	1,200	" "
1891,	275 "	2,250	" "
1894,	606 "	7,470	" "
1895 (July),	880 "	10,863	" "

CHAPTER III.

BEFORE proceeding to a consideration of purely electrical matters it will be advisable to discuss briefly the general subjects of work and activity. Suppose, for example, that a street car is being drawn at a steady rate of 5 miles an hour by a horse along a level track. Then it is evident that the horse has to do work in a mechanical sense, in order to maintain the motion. If the car could be so constructed that there was absolutely no friction in its journal bearings, and, moreover, if the road-bed could be so constructed that there were no inequalities in its metal surfaces,

15

and no friction between the wheels and the rails, then no work would have to be expended in maintaining a steady speed on a level road; or, in other words, once the car was set in motion, it would continue to run at the same rate for an indefinite period. Under practical conditions, however, as is well known, a certain amount of friction necessarily occurs and has to be overcome. The greater this friction the greater will be the amount of work which must be expended in order to keep the car running. If the road instead of being level is on a gradient, then it is evident that an ascent of this gradient necessitates the expenditure of work against gravitational force, in addition to the work expended in overcoming friction. The heavier the car and the greater its load; *i. e.*, the greater the number of passengers it carries, the greater will be the frictional work and also the gravitational work.

In order to estimate the amount of work done in any particular case, as for example, in the case above referred to of a moving car, reference is had to certain *units of work.* It is evident that when the car is being pulled by a rope, the rope is subjected to tension, such as might be produced by a weight supported over a pulley. The harder the horse pulls, the greater will be the tension and the greater the equivalent weight. Thus a horse may readily exert a pull upon its traces of 400 pounds weight. The greater the distance through which the tension is exerted, the greater will be the work done. Thus if a tension of 400 pounds weight be steadily exerted upon a car so as to draw the latter through a distance of

100 feet, then the work done will be 100
times as great as if the car were only
drawn under this tension through 1 foot;
and, generally, the amount of work, which
is performed by a tension or pull, is equal
to the tension multiplied by the distance
through which it has been exerted, so that
if the horse continues to exert a pull of
400 pounds so as to draw the car 100
feet, the horse will have expended on
the car an amount of work equal to
400 × 100 = 40,000 foot-pounds.

The *foot-pound* is not generally employed as the unit of work, except in
English-speaking countries, and, even in
such countries, scientific men generally
prefer the *joule*, a unit based on the
French system of weights and measures.
A joule is $\frac{738}{1,000}$ of a foot-pound, approxi-

mately; or, one foot-pound may be taken as equal to 1.355 joules. The foot-pound is, consequently, roughly one-third greater than the joule. If we multiply the number of foot-pounds by 1.355, we obtain the number of joules within a degree of accuracy sufficient for all ordinary purposes. For example, when a man, weighing 150 pounds, raises himself through a vertical distance of 100 feet, he performs an amount of work equal to $100 \times 150 = 15,000$ foot-pounds in the process. The same amount of work might be expressed in joules instead of in foot-pounds by multiplying the number of foot-pounds by 1.355; or, $15,000 \times 1.355 = 20,325$ joules. Again, when the horse raises a 25,000 pound car along a gradient through a total vertical distance of 100 feet, it thereby necessarily performs an amount of work against gravitation, represented by 100 × 25,000 = 2,500,000 foot-pounds. This amount of work might be expressed in joules by multiplying by 1.355 = 3,387,500 joules.

A very important distinction must be carefully kept in mind between work expended in performing any operation, and the *rate at which that work is expended;* or, as it is usually called, the *activity.* For example, a man weighing 150 pounds may raise his weight through 100 feet, by ascending a flight of stairs, in 10 minutes, or in 1 minute. The amount of work done against gravitation will in either case be the same; namely, 15,000 foot-pounds, or 20,325 joules, but it is evident that the effort which the man must exert in the two cases, and the relative degree of exhaustion which he will undergo will be very different. Ascending the flight in 10 minutes would be walking upstairs at a leisurely rate, while ascending it in 1 minute would mean running upstairs at nearly full speed. The man is obviously ten times more active in the second case than in the first; or, he expends energy ten times faster. In other words, he works ten times as fast in the second case as in the first. Consequently, activity may be defined as the rate-of-working.

The *unit of activity* generally employed in English-speaking countries is that based on the foot-pound, and is the *foot-pound-per second*, so that unit activity is the rate of expending 1 foot pound of work in 1 second. If, for example, a man raises his weight of 150 pounds through $\frac{1}{150}$th of a foot in each second of time, he expends an amount of work equal to $150 \times \dfrac{1}{150} = 1$ foot-pound in each second; or, is working at the unit rate, or with the unit activity. As this rate of working would evidently be a very small one, in dealing with large machines it is more usual to employ a unit called the *horse-power*, which is 550 foot-pounds in 1 second. Thus, when a man weighing 150 pounds, raises his weight through 100 feet in 1 minute or 60 seconds, he will perform 15,000 foot-pounds in 60 seconds, or he will average a rate of working of $\dfrac{15,000}{60} = 250$ foot-pounds per second; or, $\dfrac{250}{550} = \dfrac{5}{11}$ ths horse-power; or, will be working roughly at half the rate of a standard horse. If, however, the man ascends 100 feet in 10 minutes, he performs 15,000 foot-pounds in 600 seconds; or, at an average rate of 25 foot-pounds-per-second, that is his activity is only $\dfrac{25}{550} = \dfrac{5}{110}$ths of one horse-power.

Where the joule is employed as the unit of work, the *international unit of activity* is the *joule-per-second;* or, as it is commonly called, the *watt,* after James Watt. It is an interesting fact that James Watt introduced the term horse-power in connection with his early steam engine, and, in accordance with international usage, of naming practical units after the names of distinguished scientists, Watt's name has been selected in connection with the international unit of activity. An activity of 1 foot-pound per second is an activity of 1.355 joules-per-second or 1.355 watts. Similarly, an activity of 1 horse-power, or 550 foot-pounds-per-sec-

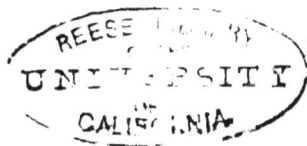

ond, is an activity of $550 \times 1.355 = 746$ joules-per-second, or 746 watts. If we multiply the number of horse-power which are being developed in any machine by 746, we obtain the activity of that machine expressed in watts. As the rate of 0.738 foot-pound-per-second is a very small unit, being about 26 per cent. smaller than the foot-pound per second, and requiring, therefore, large numbers to express large powers, in dealing with engines, it is customary to use a decimal multiple of this unit, so that the *practical international unit of activity* is the *kilowatt*, or 1,000 watts. Consequently, the horse-power, being as above mentioned 746 watts, is $\frac{746}{1,000}$ths of the larger unit, or the kilowatt, and may be taken as, approximately, 3/4ths of a kilowatt. A kilowatt will, therefore, be 4/3rds or 1 1/3rd horse-power, approximately. When we speak of a dynamo or motor as having a capacity of 100 kilowatts, (that is to say of being capable of maintaining an activity of 100 kilowatts, or 100,000 watts = 100,000 joules-per-second = 73,800 foot-pounds-per-second,) we mean an activity of 1 1/3 × 100 = 133 horse-power, approximately; or, 134 horse-power more nearly.

The problem which presents itself to the street railway manager is that of economically driving street cars by electric power, and it is to be carefully remembered that the same amount of power must be exerted by the engines in the power house as by horses drawing the cars along the streets at the same rate. In fact the engines in the power house will have to work harder, or develop a greater

activity than the horses, owing to the
necessary losses of power which inci-
dentally occur in transmission. If, for
example, we imagine that all the cars in
the streets of the city are travelling steadily
along at the same average rate as the pistons of the engines in the power house, then
the pull exerted by the pistons will be
equal to the aggregate equivalent pull of
all the cars, increased by a certain amount
corresponding to losses in transmission.

It remains now to show how power can
be calculated and expressed in electric
units. In other words, if we require to
supply a certain activity in horse-power or
kilowatts to a moving car, we need to find
how to express this power in relation to
electric circuits, since the power must be
conveyed by the electric circuits from the
power house to the car. We will, there-

fore, discuss the elementary principles of electric circuits.

An *electric circuit* is a conducting path provided for the passage of electricity. It connects an electric source or generator, with the devices to be operated by the electric current. Such a circuit is said to be *made* or *closed* when its path is completed, and is said to be *broken* or *opened* when its path is interrupted at some point or points. Thus, in the case of the electric car, an electric circuit exists between the power house where the current is generated, through the trolley wire and track, to the motors of the car. When such a circuit is closed, the current passes through the car, and drives the motor or motors. On the contrary, when the circuit is opened by the motorman at the switch, the current ceases to flow.

Fig. 1, represents a simple electric circuit consisting of a generator G, a trolley wire W W, a car with its trolley T, motors m m, and the track K K, employed as a return conductor. What passes through this circuit is an *electric flow*, generally called an *electric current*.

FIG. 1.—SIMPLE CAR CIRCUIT.

In order to obtain definite ideas concerning an electric current, a *unit of electric current*, or rate-of-flow, called an *ampere*, is employed. It will be advisable, however, before discussing the value of the ampere, to consider certain other quantities which are always intimately connected with every electric circuit. Turning our attention first to the *generator* G, it is necessary to observe that the primary function of the generator is not, as is ordinarily believed, to produce electric current, but to produce in the circuit a variety of force, called *electromotive force*, which is gener-ally abbreviated E. M. F. When the generator is driven by an engine it will supply an E. M. F. whether the electric circuit is open or closed, that is to say, whether an electric current can or cannot flow in the circuit. In other words, the generator, when running, always supplies E. M. F., but no current can be sent through the circuit until the circuit is closed. This corresponds to the case of a reservoir, which produces a water pressure whether the water be escaping under that pressure or not.

In Fig. 2, a rotary pump P, is supposed

to be placed in a power house situated by
the side of a river K K, and provided with
a pipe by which it can draw water from the
river and send it through the pipe W W.
M, is a water motor situated at some con-

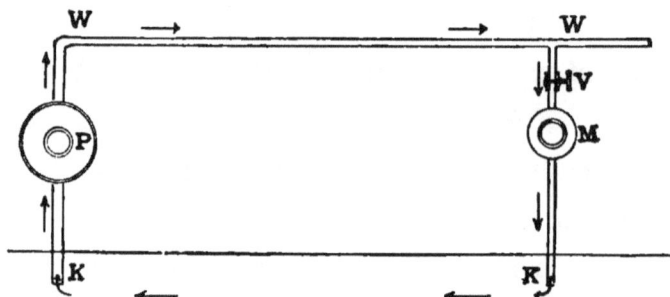

FIG. 2.—SIMPLE WATER CIRCUIT.

venient point and connected with the
main pipe W W, by a small branch pipe,
in which is placed a valve V. When the
valve is closed, the motor M, is prevented
from running, since no water current
passes through it. The *hydraulic circuit*
W W, K K, may then be said to be *broken*

or *open*.   When, however, the valve V, is
opened, water passes through the motor
M, and discharges into the river, thus *closing* the hydraulic circuit, and permitting a
water current to flow through the circuit.
It is evident that whether the valve V, be
opened or not, the generator or water
pump P, will develop, when running, a
pressure or *watermotive force* in the pipe
W W, but that no current or flow of water
can take place until the valve V, permits
it to do so, thus closing the circuit.   Here
the *watermotive force*, produced by the
action of the pump whether the hydraulic
circuit be opened or closed, corresponds to
the electromotive force produced by the
generator whether the electric circuit be
opened or closed.

The pressure generated in the supply
pipe W W, by the pump P, might be

expressed in pounds-per-square-inch; or, as
the pressure produced by a column of
water a certain number of feet in height.
In the electric circuit the pressure produced by the action of the generator G, is
expressed in *units of electromotive force*,
called *volts*. In street-car systems the electric pressure produced by the generator is
almost invariably about 500 volts; that is
to say, the pressure between the trolley
wires and the track is maintained, approximately, at 500 volts, while the pressure at
the power house between the terminals of
the generator G, may be somewhat in
excess of this, say 550 volts, in order to
make up for the loss of pressure occurring
in the circuit.

If a reservoir R, Fig. 3, filled with water
and maintained at a constant level L L, be
allowed to discharge steadily through two

pipes, as indicated in Fig. 3, one pipe A B, being a long, narrow pipe, and the other $C D$, being a short, wide pipe, it is evident that a much greater flow of water will take place in a given time through the pipe $C D$, than through the pipe $A B$, since the water pressure at the openings A

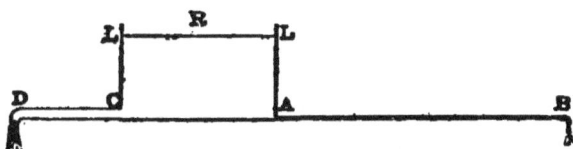

FIG. 3.—RESISTANCE OF WATER PIPES.

and C, is the same; namely, the height of water in the reservoir. The difference in the rate-of-flow of water may be ascribed to the different resistance offered by the two pipes to the flow of water, the resistance of the long, narrow pipe being comparatively great, and that of the short, wide pipe being comparatively small.

In the same way, Fig. 4, represents an electric generator *G*, which, when running, acts the part of the reservoir in the preceding case, since it supplies a steady electric pressure between its terminals. If two circuits are closed to this pressure, one through a long, thin wire *A A' B' B*, and

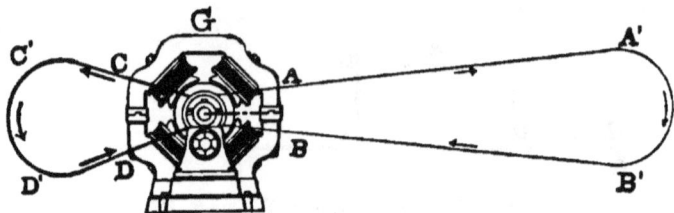

FIG. 4.—RESISTANCE OF CONDUCTING WIRES.

the other, through a short, thick wire *C C'' D' D*, then the electric flow or current, which will pass through these two circuits, will be very different, a comparatively small or feeble current passing through the long, fine-wire circuit, and a comparatively strong, or heavy current, passing through the short thick-wire circuit.

This difference in flow or current between the two circuits may be ascribed to a difference in what is called their *electric resistance*. The electric resistance of a long, thin-wire circuit is comparatively great; *i. e.*, it offers a comparatively great obstacle to the passage of electricity under the pressure of the generator *G;* while a short, thick-wire circuit has a comparatively small electric resistance; *i. e.*, it offers a lesser obstacle to the passage of electricity.

Electric resistance is usually measured in terms of a *unit of resistance* called the *ohm*, after Dr. Ohm of Berlin, who first pointed out the laws regulating the flow of electricity in conducting circuits. The amount of resistance; *i. e.*, the number of ohms in a given uniform conductor, such as a copper wire, depends upon the length of the wire, upon its area of cross-section and upon its physical condition. The longer and narrower a wire, the greater will be its electric resistance. In the same way, the longer and narrower a pipe, the greater its water resistance; on the contrary, the shorter a wire and the greater its area of cross-section, the smaller will be its resistance. An ordinary copper trolley wire, which is No. 0, American Wire Gauge, with a diameter of 0.325″, has a resistance per mile of, approximately, half an ohm, so that 2 miles of this wire would have a resistance of, approximately, 1 ohm, and 1 foot of the wire would have a resistance of $\dfrac{1}{5,280} \times \dfrac{1}{2} = \dfrac{1}{10,560}$ ohm, approximately. If the trolley wire instead of being No. 0 gauge were No. 0000, which is a wire about twice as heavy as No. 0, having a diameter of 0.46″, it

would have only half the resistance of No.
0, and, therefore, approximately, 1/4th ohm
per mile.

In Fig. 5, five copper wires, having different lengths and areas of cross-sec-

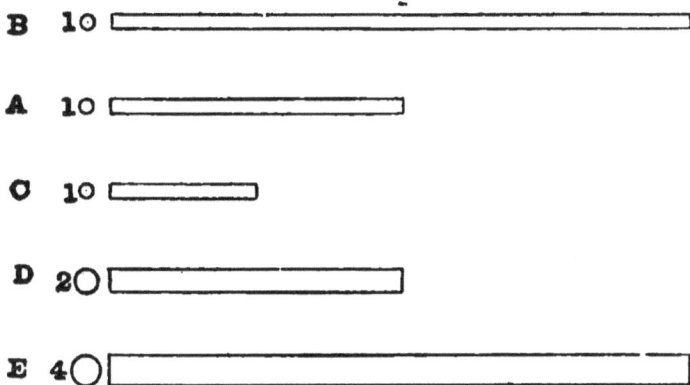

FIG. 5.—RESISTANCE OF WIRES.

tion, are diagrammatically represented.
A, represents a trolley wire 1 mile long
and 0.325″ in diameter, having, therefore, a resistance of approximately 0.5
ohm. *B*, is a wire 2 miles long of the

same cross-section, and, therefore, offering 1.0 ohm. C, is a wire half a mile
long of the same cross-section, and,
therefore, offering a resistance of, approximately, 0.25 ohm. D, is a wire 1 mile
long, but having a cross-section, as represented on the left hand side, say twice that
of any of the wires, A, B, or C. It will,
therefore, have half the resistance of A, or
$$\frac{0.5}{2} = 0.25 \text{ ohm.}$$ E, is a wire 0.65″ in
diameter, having, therefore, four times the
cross-section of A, and being 2 miles in
length. If the wire were of the same cross-
section as A, it would have $0.5 \times 2 = 1$
ohm, but being four times as heavy, its
resistance will be one-quarter of this, or
$$\frac{1.0}{4} = 0.25 \text{ ohm.}$$ Consequently, C, D,
and E, have all the same resistance,
although their dimensions are so different.

If, therefore, the cross-section and length of any copper wire be known, we can determine what its resistance will be, assuming that the conducting power of the substance of the wire is the same as that of the trolley wire we have selected as our standard. The resistance will be directly proportional to the length, and inversely proportional to the area of cross-section; or, in other words, if the length be doubled the resistance will be doubled, while if the area of cross-section be doubled the resistance will be halved.

We have hitherto considered copper wires only in estimating the resistance of a circuit. When any other conducting material, such as iron, is employed, the resistance of a wire having a given length and cross-section will be materially different. Thus, an iron wire has, approxi-

mately, 6 1/2 times as much resistance as
a wire of copper of equal dimensions.
Iron trolley wires are, therefore, never
used, for the reason that it would be necessary to employ a wire having about 6 1/2
times the cross-section of ordinary trolley
wire to have the same *conductance; i. e.,*
ability to conduct electric current. Iron,
however, enters into street railway circuits
in the form of the tracks, which, as we
have seen, form a portion of the return
circuit to the power house.

The dimensions of a wire which has a
resistance of 1 ohm will necessarily vary
with the character of the material of which
the wire is composed. Thus, in copper, its
length might be approximately 2 miles, if
its diameter was that of a trolley wire,
0.325"; or, its length might be only 1
foot, if of No. 40 American Wire Gauge,

having a diameter of 0.003145″; if of iron, a length of about 900 feet of trolley wire; and, roughly, 2 inches of No. 40 wire would have a resistance of 1 ohm. In all cases the exact resistance would depend upon the degree of purity of the metal, as well as upon its physical condition; that is to say, upon its hardness, and temperature. Since mercury is a metal, which is fluid at ordinary temperatures, and can be readily obtained in a nearly homogeneous and pure condition, the ohm has been practically defined as the resistance of a column of mercury 1.063 metres in length, and 1 square millimetre in cross-section, at the temperature of melting ice.

It is evident from what we have said that the quantity of water which flows, in any given time, through the pipe referred to in connection with Fig. 3, will depend both on the pressure or head of water in the reservoir, as well as upon the resistance which the pipe offers to the flow. In the case of the electric circuit the same rule applies, that is to say, the *quantity of electricity* which passes or flows in an electric circuit, depends not only upon the electric pressure in the circuit which causes the flow, but also upon the resistance of the circuit which opposes it.

In the case of the electric circuit the electric current is related to the E. M. F. and to the resistance in accordance with a law generally known as *Ohm's law*. This law may be expressed as follows:

The current strength in amperes flowing through a circuit, varies directly with the pressure or E. M. F., and inversely with the resistance; so that if we divide the number of volts in the E. M. F. by the number of ohms in the resistance, we obtain the current strength in amperes; or, concisely, amperes $= \dfrac{\text{volts}}{\text{ohms}}$.

Thus, if a circuit contains an E. M. F. of 10 volts, and a resistance of 5 ohms, the current in the circuit would be $\dfrac{10}{5} = 2$ amperes.

We have seen, in connection with Fig. 3, that the quantity of water which flows per second through the water pipe from the reservoir, depends both on the pressure at the reservoir, and on the resistance of the pipe. This, however, is only true when no obstacle to the flow of the water exists save the resistance of the pipe itself. If, for example, instead of permitting the water

to escape freely from the open end of the
pipe it be first caused to pass through, and
actuate, a water motor, then the conditions of flow will be profoundly modified,
much less water flowing through the pipe
in the second case than in the first.   If, for

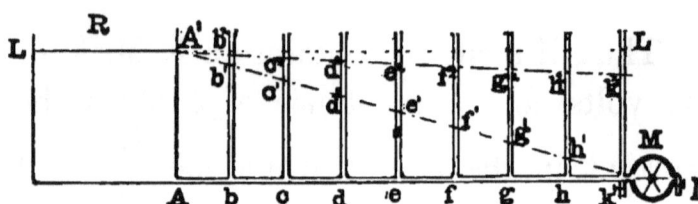

FIG. 6.—HYDRAULIC GRADIENT.

example, as in Fig. 6, the reservoir R, is
capable of discharging by the pipe A k',
either through the faucet k', into the air,
or through the faucet l, after passing
through the motor M, the flow in the two
cases will be very different.   In the first
case the available pressure at the reservoir will be
that due to the height of the water A A',
say 50 feet, while the pressure at the dis-

charge point, will simply be that of the
external air, or a column of 0 feet. In
other words in discharging through the
pipe the water pressure suffers a drop as
represented by the dotted line $A'\,k'$, and
the pressure at the intermediate points is
indicated by the points b', c', d', e', f', g', h'.
If, however, the faucet k', be closed, and
that at l, be opened, thereby establishing
communication through the water motor
M, the motor will commence to operate,
and in so doing will develop a *back pressure*, or *counter watermotive force*, which
opposes the flow of water and acts like a
resistance. The pressure at k', under these
circumstances, instead of being 0 feet, will
rise to k^2, and the drop of pressure, which
has taken place in the tube $A\,k'$, will have
diminished from $A\,A'$ to $k^2\,L$, with a correspondingly reduced flow of water through
the pipe.

Similarly, if the electric circuit represented in Fig. 2, be so modified as in Fig.
7, that it may be closed either at *c c*, directly back through the track, or at *H,*

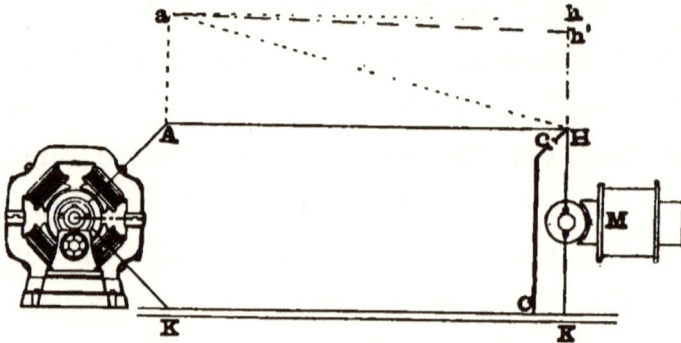

FIG. 7.—ELECTRIC GRADIENT.

through an electric motor *M,* the electric
flow or current in amperes will be very
different in the two cases. If the circuit
be closed through the track wire at *c c,*
the pressure, at *A,* will be say 500 volts,
as represented by the dotted line *A a,* and
supposing the length *A H,* to be 1 mile
of trolley wire, then neglecting, for con-

venience, the resistance of the track and generator, the resistance of the circuit will be 0.5 ohm, and the current strength in the circuit 500 volts ÷ 0.5 ohm = 1,000 amperes.

If, however, the circuit be closed through the motor M, the latter will be actuated by the current and will be set into rotation, whereby a *back pressure*, or *counter electro-motive force*, usually abbreviated C. E. M. F., will be set up in the motor, of say, 450 volts, as represented by the dotted line $H\,h'$; so that the *effective pressure or E. M. F.* which drives the current through the circuit, will be reduced to $h\,h' = 500 - 450 = 50$ volts, and the current strength, neglecting the resistance of the generator motor and track, will be, 50 volts ÷ 0.5 ohm = 100 amperes.

A flow of water is sometimes rated as being a certain quantity of water; *i. e.*, a certain number of cubic feet or gallons per second. In the same way the electric flow may be rated as being a certain quantity of electricity passing through the circuit per second. The unit of electric quantity is called the *coulomb*, and has been so chosen that a flow of 1 coulomb per second is called an ampere. Consequently, a flow or current of 1 ampere, maintained in a circuit for 1 minute, represents a total flow of 60 coulombs of electricity, and, maintained for one hour, a total flow of 3,600 coulombs.

When an E. M. F. acts on a broken or open circuit, it is unable to send any current through the circuit, and will, therefore, do no work. Thus, when the generator at the power house is driven by an engine and supplies an E. M. F. of 500 volts to the trolley system connected with it, no current will pass through the generator if there be no cars on the line, assuming that the wires are properly insulated. Under these circumstances the generator will not be supplying any power, and the engine will have no work to do except to drive the generator against its friction. In fact, except that the generator armature is magnetized, it behaves like a mere wheel of copper and iron, so supported on an axis in bearings, that it might be rotated with a very small expenditure of power. When, however, the circuit of the generator is closed by the connection of the cars with the trolley wire, so that a current is transmitted through the circuit or circuits under the pressure of 500 volts, the generator does work at a rate which will depend upon the amount of current supplied, the greater the current strength in amperes delivered to the trolley system, and distributed to the cars, the greater will be the activity which the generator has to supply, and the greater will be the activity which the engine must supply to drive it, so that when the load comes on the system by the operation of the cars, the generator which previously required say 20 horse-power only to revolve it, may now require the engine to supply 500 horse-power, which activity will be transformed into electric activity in the circuit. If we multiply the pressure in volts by the current strength in amperes which is being supplied by that pressure, we obtain the activity supplied in watts. Thus, if a generator supplying 550 volts at its terminals to a trolley system delivers a current strength of 50 amperes through the circuit containing its armature, trolley, street-car motor, and track, then the activity supplied by the generator at its terminals will be 550 volts × 50 amperes = 27,500 watts = 27.5 kilowatts (usually abbreviated KW) = 36.85 HP = 27,500 joules-per-second = 20,268 foot-pounds-per-second. The engine would have to supply more power than this to the generator, since it would have to make up for the loss of power in the generator owing to its mechanical and electrical frictions, but if the generator had an efficiency of 90 per cent., that is to say, if its output was 90 per cent. of its intake, then the activity which the engine would have to supply to the generator would be $\frac{27,500 \times 100}{90}$ = 30,555 watts = 30.555 KW = 40.94 HP = 30,555 joules-per-second = 22,517 foot-pounds-per-second.

Just as the total amount of work expended by water escaping from a reservoir, is equal, in foot-pounds, to the number of pounds of water multiplied by the number of feet through which it falls, so the total amount of work expended by electricity in flowing through a conductor or circuit is equal, in joules, to the number of coulombs of electricity multiplied by the number of volts difference of electric level, or pressure, under which it passes. Thus a current of 50 amperes flowing under a pressure of 550 volts, represents a flow of 50 coulombs-per-second under that pressure and an amount of work equal to $50 \times 550 = 27,500$ joules in each second, or, in one hour of 3,600 seconds, a total work of $3,600 \times 27,500 = 99,000,000$ joules. But we have seen that the activity in this circuit is 27,500 watts, and this activity maintained for an hour will require an expenditure of 27,500 watt-hours, or 27.5 kilowatt-hours. A *watt-hour* is, therefore, a quantity of work equal to 3,600 joules, or 2,657 foot-pounds, while a *kilowatt-hour*, the unit of work usually employed with large electric machines, will be 1,000 times as much, or 3,600,000 joules = 2,657,000 foot-pounds.

If a pressure of 550 volts is maintained steadily at the generator terminals, under all conditions of load, the pressure at the trolley of the single car we have considered, will be less than 500 volts by an amount which will depend upon the size and number of the conductors in the network supplying it, and upon the length of those conductors, or the distance of the car from the power house. Thus, if the car be 1 mile from the power house, and if the track have, for simplicity, a negligi-

ble resistance, while the single trolley wire
supplying the car has a resistance of 0.5
ohm per mile, then the resistance between
the generator and the car will be 0.5
ohm, and the drop in this length of conductor will be 50 amperes × 0.5 ohm =
25 volts, so that the pressure at the terminals of the car motor as determined by
a *voltmeter*, or instrument for measuring
the number of volts, would be 550 − 25 =
525 volts, and when the car was operating,
the voltmeter, if connected between the
trolley wire and the track at the car,
would show this pressure, while as soon as
the car was disconnected by opening the
switch, the pressure between the trolley
wire and the track would immediately
rise to 550 volts, assuming no other car or
leakage current to exist over the system.
The amount of drop which will be produced over a given length of conductor

will depend entirely upon the current strength, so that if we double the current strength we double the drop.

The activity which the motor will receive at its terminals will be the current strength in amperes, (which is the same all through the circuit when only one car is employed,) multiplied by the pressure at its terminals. Thus, in the preceding case, the pressure being 525 volts at the motor terminals between trolley and track, while the current strength is 50 amperes, the activity absorbed by the motor will be 525 volts × 50 amperes = 26.25 KW, or 1.25 KW less than that supplied by the generator to the line. This activity of 1.25 KW is expended in the line as heat, uniformly distributed through its substance; for, the drop being 25 volts, and the current strength 50 amperes, the activity expended in this conductor will be 25 volts × 50 amperes = 1,250 watts, = 1.25 KW expended entirely as heat.

Of the 26.25 KW delivered to the motor, only a certain fraction will be usefully employed in driving the car, the remainder being uselessly expended in heating the motor. If the efficiency of the motor be 80 per cent., then the activity usefully expended in the preceding case will be $26.25 \times \dfrac{80}{100} = 21$ KW = 28.14 HP = 21,000 joules-per-second = 15,480 foot-pounds-per-second. This activity will be supplied to the shaft of the motor. Assuming at present that no power is wasted in gears, then this activity will be available for propelling the car. For example, if the car friction were very small, and its total weight, including passengers was 30,000 pounds, then the activity supplied would be capable of lifting 30,000 pounds through a distance of $\frac{15,480}{30,000} = 0.516$ foot-per-second.   With a 1 per cent. grade this would represent a speed of 51.6 feet-per-second, or 35.2 miles-per-hour, and with a 10 per cent. grade it would represent a speed of 5.16 feet per second, or 3.52 miles-per-hour.

It is evident, therefore, that the activity which can be communicated to a moving car for a given activity supplied at the driving shaft of the engines, depends upon the efficiency of the generator, the efficiency of the motor, and the efficiency of the line conductor, including under this term, the track.

The *efficiency* of a motor or generator is

the ratio of the output to the intake. The
efficiency of a line conductor or circuit
may also be regarded as the ratio of the
output to the intake, the intake being
measured at the generator terminals and
the output at the motor terminals. The
efficiency of a generator or a motor usually
increases with the load up to full load or
nearly full load, so that, under ordinary
circumstances the more work we can get
the motor or generator to do, within the
limits of its capacity, the greater the proportion of useful work delivered, to the work
received, although the loss of work will
be absolutely greater. Thus, a street car
motor, whose maximum activity is rated
at 15 KW (approximately 20 HP) would
require, perhaps, 2 KW to run it when
entirely free from all load or disconnected
from its gears; *i. e.*, when doing no useful work, so that its efficiency would be

$\dfrac{0}{2} = 0$. When fully loaded, however, it might waste 3 KW and deliver 15 KW, so that its intake would be 18 KW, and its efficiency $\dfrac{15}{18} = 0.833 = 83.3$ per cent. Its efficiency may, therefore, increase from 0 to 83.3 per cent. from no load to full load, although the actual loss of activity in it would increase in the same range from 2 KW to 3 KW. The same principles apply to a generator, and for this reason it is always more economical to operate generators at a fair proportion of their full load.

In the case of the line conductor or conductors, including. track conductors, the case is different. The efficiency is always less as the load increases. Thus, if we supply a current strength of 1 ampere over a circuit of trolley conductor and

track, having a total resistance of 1 ohm,
then the drop in this circuit will be
1 ampere × 1 ohm = 1 volt, and if the
pressure at the motor be kept at 500 volts,
the pressure at the generator will have to
be adjusted to 501 volts; or, if the pressure
at the generator be kept at 500 volts, the
pressure at the motor terminals will, with
a current of 1 ampere, automatically become 499 volts. If, however, 2 amperes
be supplied through the same circuit,
the drop will double, or will become 2
volts, and the pressure at the generator
will be 502 volts, if that at the motor is
500. In the former case the efficiency of
the line circuit will be $\frac{500}{501}$; in the latter
case it will be $\frac{500}{502}$. Similarly, if the current strength be increased to 100 amperes,
the drop will increase to 100 volts, and

with 500 volts at the generator there will
be 400 volts left at the motor, making the
efficiency $\frac{400}{500} = 0.8 = 80$ per cent. It is
evident, therefore, that the efficiency of the
line continuously decreases with the load.

It is clear from the preceding that if
a trolley wire were very long, say 15
miles, so that its resistance was 7.5 ohms,
then the current strength of 50 amperes
passing through the circuit to operate the
car motor at the extreme distance from
the power house would produce a drop of
50 amperes × 7.5 ohms = 375 volts, leaving
only 175 volts pressure at the motor when
550 volts was the pressure at the generator
terminals, and assuming no resistance in
the ground-return circuit. The activity
delivered by the generator would be 550
volts × 50 amperes = 37.5 KW. The

activity available at the motor terminals
would only be 175 volts × 50 amperes =
8.75 KW, so that the efficiency of the line
would only be $\frac{8.75}{37.5}$ = 0.319 = 31.9 per
cent., while the available speed of the car
would be correspondingly reduced.   In
other words, owing to the great length of
conductor, and resistance in the circuit, a
large percentage of the activity would be
expended in heating a long length of wire,
instead of driving the car.

The same condition of line efficiency
would be produced by a number of cars
over a shorter length of circuit.   Thus,
reverting to the case of a single mile of
trolley wire, if a bunch of five trolley cars
should start together from the distant end
of the line towards the power house, each
taking 50 amperes of current strength, the

total current strength supplied to the bunch would be 250 amperes, and the drop in the line would be 250 amperes × 0.5 ohm = 125 volts, making the pressure at the bunch 425 volts. The line efficiency, under these conditions would be $\dfrac{425}{550}$ = 0.772 = 77.2 per cent. Consequently, when the distance to which cars have to be run is great, or, when the number of cars and the current strength to be collectively supplied are great, the amount of copper employed to supply the system must be increased so as to reduce the effective conductor resistance. If, for example, we double the area of cross-section of the trolley wire, and, therefore, its weight per mile, we halve the resistance of the conductor per mile and, consequently, halve the drop, excluding track resistance, and, therefore halve the drop which will occur at any given distance with any given load.

There is, however, an obvious limit to the size of trolley wire which can be practically employed. In fact, trolley wires are almost always constructed of No. 0, A. W. G. They are supplemented, however, in practice, by what are called *feeders;* *i. e.*, feeding conductors which are separate from the trolley wires, but which lead from the generator in the power house and connect with the trolley wire at suitable distances along the track. Thus in Fig. 8, G, is the generator, and C, a car at a certain distance along the track. $G F_1$, $G F_2$, $G F_3$, $G F_4$, four separate feeders connecting with the trolley wire at different distances. As shown in the diagram, the current strength required to supply the car, is probably supplied

in a large measure by feeder $G\ F_1$, so that
the feeders $G\ F_2$, $G\ F_3$, and $G\ F_4$, are
comparatively idle. Consequently, the drop
of the feeder $G\ F_1$, will be comparatively

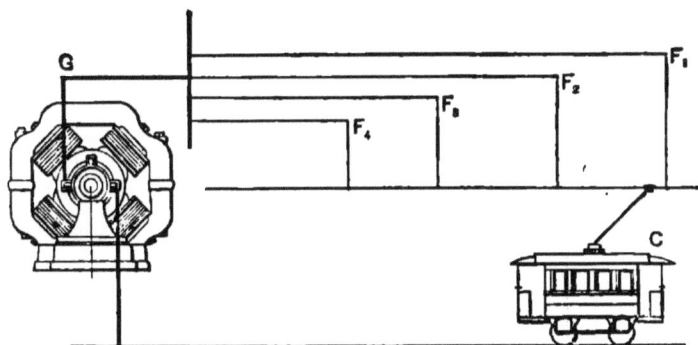

FIG. 8.—FEEDER SYSTEM.

great with reference to that of the other
feeders. F_1, F_2, F_3, and F_4, are called *feeding points*.

In practice it is usual to so arrange the
feeders and the distances between feeding
points, that when all the cars are being

operated at average distances, the drop shall nowhere be in excess of 50 volts, and, therefore, that with 550 volts at the generator terminals the pressure shall not be lower than 500 volts at any point on the line.

CHAPTER IV.

THE MOTOR.

As is well known, the power which propels a trolley car is obtained from the electric current transmitted through the circuit, by the intervention of an *electric motor* or motors, there being usually two motors placed on the truck of an ordinary street car. Fig. 9, shows the general construction of a truck with two motors M, M, in place, one geared to the axle of each pair of wheels. Reserving for description in Chapter V. the different methods adopted for the mounting or hanging of a motor, as well as the details in the construction of the car truck, we will now proceed to the general description of the motor, its construction and operation.

Fig. 10 shows a form of electric motor in extended use. Here the motor is com-

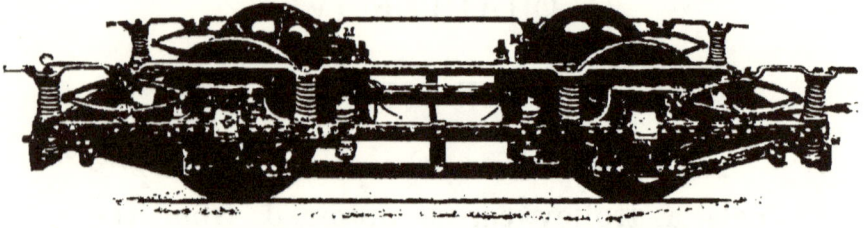

FIG. 9.—CAR TRUCK WITH MOTORS IN PLACE.

pletely enclosed in a cast-steel frame F, F, F, made in two halves, fitted together, as shown. Since the motor runs within a few inches of the surface of the street, and is, therefore, exposed to dust, mud and water, it becomes absolutely necessary not only to provide it with a casing, but also to make this casing practically air and water tight. The main shaft of the motor is seen projecting through its bearing at *A*, and this bearing is lubricated by the grease box *C*.

FIG. 10.—FORM OF ELECTRIC MOTOR.

The armature shaft is connected with the axle of the wheels on which the truck rests, by *gear wheels* enclosed in the gear cover *G, G*. The gears are inserted in order to reduce the speed of the car as well as to increase the effective pull of the motor, as will be more clearly pointed out subsequently. The main axle passes through the bearing B, lubricated by the grease box C'. The motor is supported on the truck by the lugs L', L'. Access to the working parts of the motor is had by the lid L, L, L, while a more nearly complete inspection can be obtained by unscrewing two bolts, one of which is seen at B, and throwing back the upper half of the motor upon hinges H, H. The insulated cables K, K, pass through holes in the castings and supply electric current to the motor. This particular motor is called a G. E. 800 motor, the number 800 representing that it is capable of exerting on the car a push of 800 lbs. weight at the main axle, when supplied with the full current strength, and mounted on 33″ wheels

on level rails. Two such motors when supplied with full current strength, therefore,
give a push of 1,600 lbs. weight to a car.

Fig. 11, shows the same motor with the
upper half thrown back on its hinges, thus
permitting an inspection of the parts of
the motor. Here, as in all this class of
electric motors, the essential parts consist
of an armature or rotating part A A, with
a commutator at M M, upon which rest
the brushes C, C, which carry the current
from the trolley line into and out of the
armature. The armature rotates between
four poles, of which one is shown in the
upper lid at P, surrounded by a magnetizing coil of wire W. The armature shaft
has a pinion N, secured to one of its extremities, which engages with a gear-wheel
on the main axle of the truck, which axle
passes through the bearings B, B.

The armature of one of the electric motors above described consists essentially of three parts; namely, the *armature*

FIG. 11.—MOTOR OF FIG. 10 OPENED.

core, mounted on its shaft, the *armature windings* or *coils*, which are placed on the armature, and the *commutator*. The general appearance presented by an armature core, mounted on its shaft, is shown in Fig. 12. Here, as will be seen from an

Fig. 12.—Unwound Armature.

inspection of the figure, the core consists of a cylindrical body made of soft iron. If the armature core be made from a

solid mass of iron, it has been found by experience that during the changes in mag-
netization to which it is subjected, when it
rotates, deleterious electric currents called
eddy currents, are generated in it. These
currents cannot be employed in the external circuit; they merely serve to heat
the armature core and so prevent the efficient operation of the motor. By adopt-
ing the simple expedient of *laminating* the
core; that is, of forming it of thin sheets
of iron, laid side by side, this difficulty is
avoided. The armature core shown in
Fig. 12, is laminated, that is, formed of
discs or rings clamped together and supported at right angles to the axis of the
shaft. The edges of the cylindrical iron
core thus formed are provided, circum-
ferentially, with a series of longitudinal
grooves or recesses, as shown. These are
intended for the reception of the insulated

copper conductors that carry the electric current.

In placing the insulated copper wire on the armature core, care is necessary to obtain a symmetrical disposition of the wires. One method of arranging the conductors on the core is shown in Fig. 13, which represents an armature in the process of winding. Armatures for motors are made in a variety of forms of which, perhaps, the *ring armature* and the *cylinder armature* are the commonest. The armature shown in Fig. 13 is of the cylinder type. Here the wire is wound only on the outside of the core. A single cotton-covered wire, starting at say *A*, passes to *B*, through the grooves, provided on the surface of the core for its reception. It then descends to *C*, in the curved path shown, turns inwards and passes on to *D*, when it

again crosses through the groove to *E*, and
so on.   All the wires which are left projecting on the left-hand side are intended

FIG. 13.—ARMATURE IN PROCESS OF WINDING.

to be connected to the part called the commutator, the object of which will be explained subsequently.

A particular form of commutator is
shown in Fig. 14.   It consists, as shown,

of a number of segments of copper placed
longitudinally on the surface of a cylinder,
each strip being insulated from the adja-

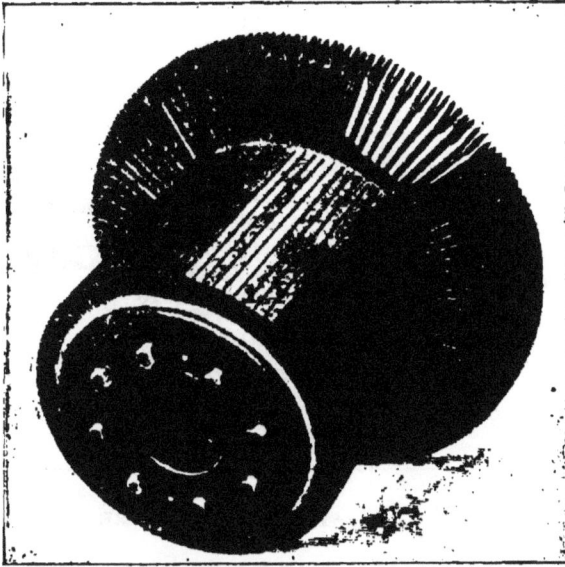

FIG. 14.—FORM OF COMMUTATOR.

cent strips by means of a thin plate of
mica. The *commutator strips, segments,* or
bars, as they are called, are connected to
the free ends of the wires which are

soldered into the clips left for them. Fig. 15, shows a completed armature, or the appearance of the armature in Figs. 12 and

FIG. 15.—WOUND ARMATURE.

13, when the process of connecting and soldering is complete.

It now remains to explain the manner in which the electric current passing through the armature causes it to rotate. When the current enters the armature conductors at one brush and circulates around the coils of wire wrapped on its surface, it also passes through the coils of wire around the field magnets. By these means both the armature and the field poles are rendered magnetic, and it is to the magnetic attractions and repulsions that take place between the movable armature and the fixed field poles, that the rotation of the armature and the mechanical force it develops are due. Since, however, the form of electric motor employed in the street car is very compact and difficult to understand, it will be preferable first to consider a few simpler types of electric motors.

It is a well known fact that when two magnets are brought near together, their unlike poles; *i. e.*, the north pole of one

and the south pole of the other, will
attract, while their like poles will repel,
so that if one of the magnets be free to
move, it will come to rest in such a posi-

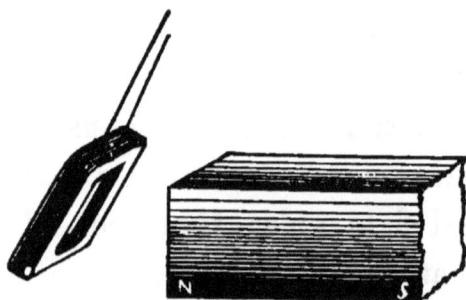

Fig. 16.—Action Between Magnet and Active Coil.

tion that opposite poles are adjacent. A
conductor carrying an electric current,
acts like a magnet, so that if a magnet be
approached to an *active coil* of conductor;
i. e., a coil carrying a current, as shown in
Fig. 16, an attraction will take place between the unlike pole of the magnet and
the active coil. In the case of the coil of

insulated wire, shown in Fig. 16, the faces of the coil become magnetic, as marked at *S* and *N*. If the direction of the current through the coil be reversed, the polarity of the coil will be reversed, so that, if the coil were free to move, it would turn around and present its opposite end to the magnet; or, if prevented from doing this, would be repelled bodily by the magnet.

If now the coil, instead of being sus-pended by the two wires which carry the current into and out of it, is placed as shown in Fig. 17, that is, suspended flat and horizontally in the position *a b c d*, by the two wires before the north pole *N*, of the bar magnet, then, as soon as a sufficiently powerful current is passed through the coil, it will set itself at right angles to the magnet into the posi-tion *a' b' c' d'*, as shown by the dotted

lines. If the current through the coil be
reversed, the coil will turn around and
present its opposite face to the magnet.
This action can be intensified by employ-

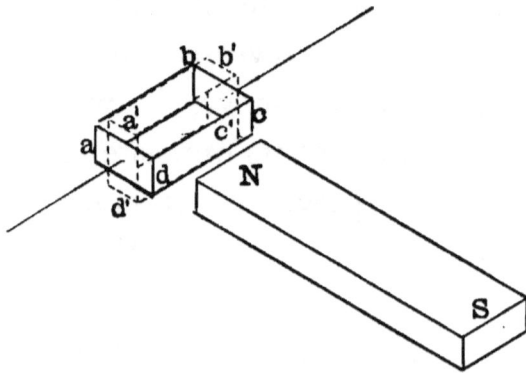

FIG. 17.—DEFLECTION OF ACTIVE COIL BY MAGNET.

ing two bar magnets with opposite poles
at *N* and *S*, as shown in Fig. 18; for, each
magnet attracts the opposite face of the
coil. By combining the two bar magnets
into a single horseshoe magnet in the
manner shown in Fig. 19, the action on

the coil can be rendered still more powerful.

In the simple form of apparatus shown
in Figs. 17 to 19, the coil has been sup-

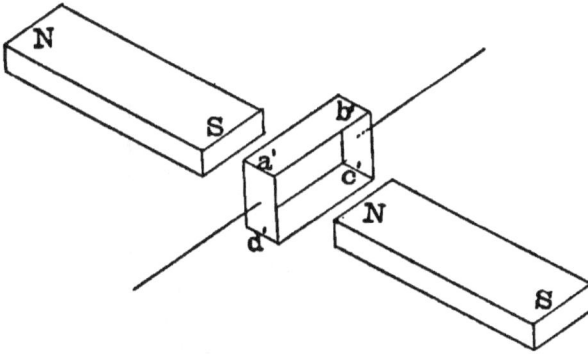

FIG. 18.—DEFLECTION OF ACTIVE COIL BY OPPOSITE
POLES OF TWO MAGNETS.

ported in air. If, however, the coil be
wound upon a cylinder of iron, as shown
in Fig. 20, the magnetic power with which
it tends to rotate is very much increased.
Moreover, instead of employing a *permanent horseshoe magnet*, we may wind a coil

of insulated wire *C C*, around the *soft iron horseshoe magnet core*, shown in Fig. 20, and by passing an electric current through this wire we may obtain a more powerful

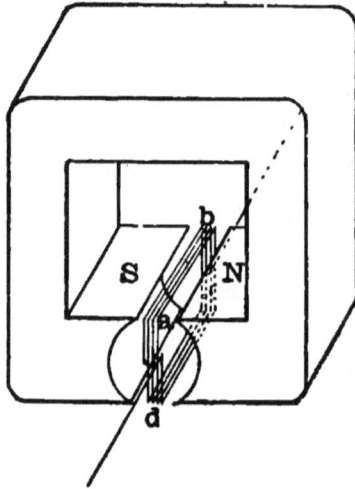

FIG. 19.—DEFLECTION OF ACTIVE COIL BY HORSESHOE MAGNET.

magnet than would be possible with any permanent magnet of steel. By this means we obtain a still more powerful *electromagnetic twist* or *pull*, technically called the *torque*, when the current is allowed to pass through the armature coil.

It is evident that in the preceding cases

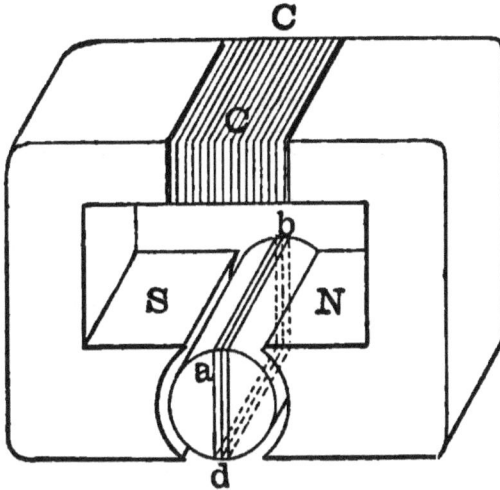

FIG. 20.—DEFLECTION OF ACTIVE COIL WOUND ON IRON CORE BY ELECTROMAGNET.

the motion of the coil will cease as soon as it sets itself at right angles to the line joining the magnetic poles. If, however, the current in the coil could be automatically reversed; *i. e.*, changed in direction, as soon as this position was reached, the armature would turn round, or rotate, through half a revolution, when it would again come to rest at right angles to the line joining the poles *N, S.* The device, whereby the direction of the current through the coils is automatically reversed every time that the coil sets itself in the neutral or dead position, so as to ensure another half rotation, is called a *commutator*, because it commutes or changes the direction of current in the coils at the desired moment.

Early forms of electric motors employed only a single coil on the armature, as represented in Fig. 20, but later forms invariably employ a number of coils disposed at uniform angular distances around the surface of the armature so as to main-

tain the twisting power or torque uniform in all positions.

The continuous-current electric motor,

FIG. 21.—STATIONARY ELECTRIC MOTOR.

as in actual use on street cars, consists substantially of a suitable combination of the
parts just described; namely, of the arma-

ture, of the field magnets and their poles, and of the commutator. A practical form of stationary electric motor is shown in Fig. 21, where *N* and *S*, are the poles of a powerful electromagnet wound with many turns of insulated wire, and *A*, the armature, which rotates between these poles. *C*, is the commutator upon which the brushes *B*, *B*, rest in such a manner that, by the rotation of the armature, the direction of current in the loops of wire is changed at the moment required to ensure a continuous rotation.

Motors are made in a great variety of forms. For example, instead of having only two poles, four or more poles may be employed. Thus Fig. 22, shows a form of *four-pole* or *quadripolar motor*, with its four magnetizing coils *N*, *S*, *N*, *S*, provided to produce the four poles. In this particular case four sets of brushes B, B, are employed, of which only three are visible in the cut. The armature A,

FIG. 22.—STATIONARY QUADRIPOLAR MOTOR.

revolves in the space between the four poles, and the current is supplied to this armature from the brushes B, B, through the commutator M. Here the field frame $F\,F\,F$, is of cast iron.

Street-car motors are almost always of the quadripolar type. Owing to the fact that these motors have a very small space allotted them under the car, and are required to be very light, the four magnet poles are as short as possible, and the field frame, instead of being made of cast iron, is of soft cast steel, which is much more advantageous from a magnetic point of view. In the motor of Fig. 11, there are four poles, two only of which, the upper and lower, are wound with coils of wire. The poles on the side being unwound or being, as they are sometimes called, *consequent magnetic poles*. Fig. 23, shows the castings for another form of quadripolar street-car motor. In this case, each of the four poles N, S, N, S, is surrounded by a

magnetizing coil, and the whole field frame
$F F F$, is of cast steel. In order to permit
access to the interior of the field frame, it

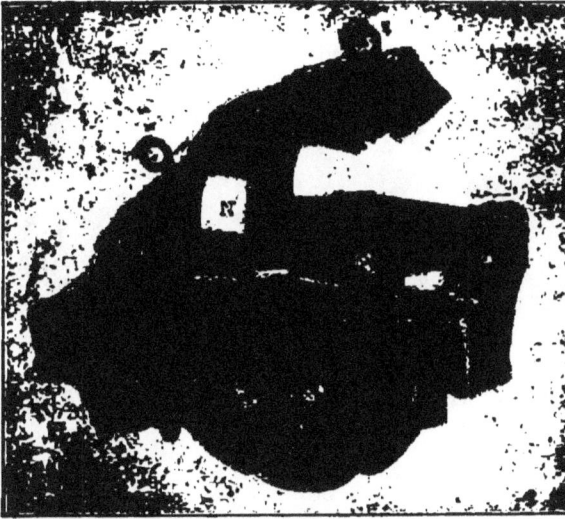

FIG. 23,—FIELD-FRAME CASTINGS OF QUADRIPOLAR
STREET-CAR MOTOR.

is made in halves and the upper is movable
on a hinge P. The armature for this
motor is shown in Fig. 24 in three successive conditions. At A, is seen the un-

wound core composed of sheets of iron
punched with radial teeth, so as to form,
when assembled, a compact cylinder with
grooves or slots as shown. At *B*, the

FIG. 24.—ARMATURE FOR MOTOR OF FIELD FRAME IN
FIG. 23.

insulated conductors have been placed in
these grooves ready for connection to the
commutator at the distant end of the core,
while at *C*, the finished armature is shown.
The appearance of a similar motor, after
being assembled, is shown in Fig. 25.
Here *A*, is the armature geared to the main

axle through reducing gear, covered by the gear cover *G G*. *B, B,* are two brushes,

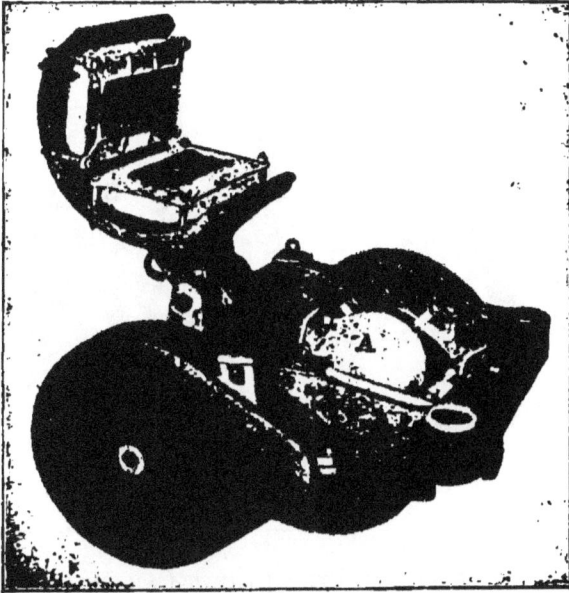

FIG. 25.—ASSEMBLED MOTOR OPEN FOR INSPECTION.

the armature winding being such that only two brushes need to be employed. This is the plan generally adopted with street-car motors, while stationary quadripolar machines usually employ four brushes or sets of brushes, as shown in Fig. 22. *N, S,* are the two poles in the upper half of the field frame, each being surrounded by a

FIG. 26.—COMPLETED STREET-CAR MOTOR.

magnetizing coil. The completed motor, closed and ready for suspension, is shown in Fig. 26. Here *B,* shows one set of brushes protected from dust and mud by the shell *S.* *F F,* is the field frame, *G G,* the gear cover. *A*, the armature shaft and *R*, the truck-wheel shaft. *C, C, C,* the terminals of the motor from which wires lead to the controller or car switch.

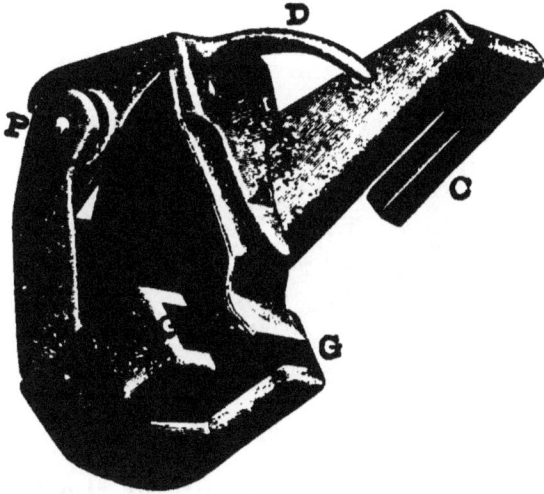

FIG. 27.—BRUSH HOLDER.

A form of *brush holder* employed in the motor of Fig. 11, is shown in Fig. 27. This brush holder is of metal and is clamped in the slot *C,* to its supporting frame through which it receives the electric current. The brush slides freely in the guides *G*, *G*. The brush being composed of a rectangular block of carbon, the arm *A*, pivoted at *P*, maintains a uniform pressure

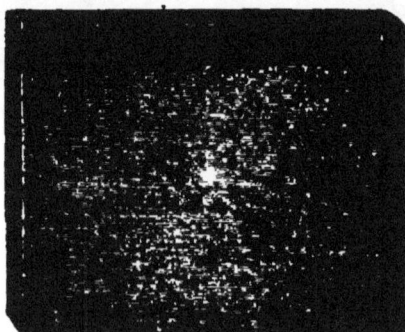

FIG. 28.—CARBON BRUSH.

at the back of the brush under the tension of the spiral spring *S*, thus pressing the brush against the surface of the commutator beneath. The arm *A*, can be withdrawn, and the brush lifted, by pulling with the finger upon the tongue *D*. A form of such brush is shown in Fig. 28.

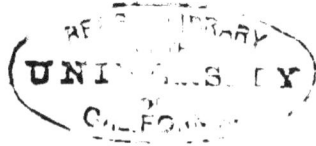

CHAPTER V.

CARS AND CAR TRUCKS.

A STREET car, as it appears on the street, is composed of two distinct parts; namely, the *car body*, or the enclosed space for the passengers, and the *car truck*, or the part upon which the car body rests. Limiting our present consideration to the car truck, we find that this consists generally of a frame resting upon the axles of the wheels, through journal boxes.

There are three methods of supporting car bodies on trucks; viz.,

(1) By the use of a single rigid truck with four wheels and two axles, the axles

remaining sensibly parallel in all positions
of the car, whether on curves or on straight
tracks.

(2) By the use of two trucks, one at
each end of the car. In this case the car is
usually supported upon the swivel centre
of each truck.

(3) By the use of three trucks, the car
being supported on the end trucks, and
the centre truck being movable, so that the
car axles are only parallel on straight
tracks, and are radial on curves.

A single truck is commonly used for
short cars and the double or triple truck
for long cars.

Fig. 29, shows a particular form of
single truck. *F*, *F*, *F*, are solid forged
side frames. *B*, *B*, are the journal boxes,
in which the axles run, and on which the

weight of the car rests, through the double spiral springs *S, S.* The car body is supported on the steel beams *B', B',* which, in their turn, rest upon the side frames through the four spiral springs, and the

FIG. 29.—SINGLE TRUCK.

two elliptical springs on each side. The wheels are provided with *brake shoes L, L.*

A form of truck for a *double-truck car* is shown in Fig. 30. Here the motor is mounted so as to drive the left-hand axle, and the weight of the car is so disposed upon the truck as to throw the principal share of the weight upon this pair of wheels in order to provide sufficient traction and prevent the rotation of the motor from causing the wheels to slip.

Fig. 31, shows another form of truck for a double-truck car called a *maximum trac-*

FIG. 30.—TRUCK OF DOUBLE-TRUCK CAR.

tion truck. This truck has two axles, and two pairs of wheels of different diameters. The motor is suspended in such a manner as to drive the larger pair, nearly 9/10ths of the weight of the car being distributed upon these wheels so as to obtain the maximum tractive effort.

Fig. 32, shows a *triple-truck support*, called a *Robinson radial truck*. Here the

FIG. 31.—MAXIMUM TRACTION TRUCK.

car is supported upon the centres of the end trucks in such a manner that these may swivel freely, carrying the middle truck between them. Fig. 33 illustrates the action of these trucks when going around a curve. It will be seen that the middle truck is pulled over to that side of

the car body which is on the outside of
the curve. The advantage of double and

FIG. 32.—ROBINSON RADIAL TRUCK.

triple trucks is considerable with long cars,
but for short cars they are usually considered unnecessary, although they save
some power and wear going around curves.

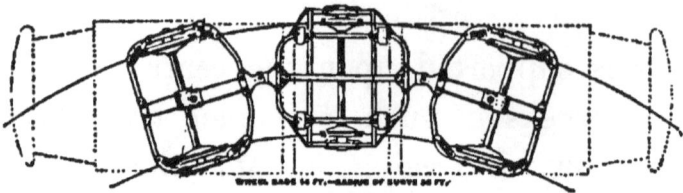

FIG. 33.—ACTION OF RADIAL TRUCK.

The appearance presented by a single-
truck car is illustrated in Fig. 34, which

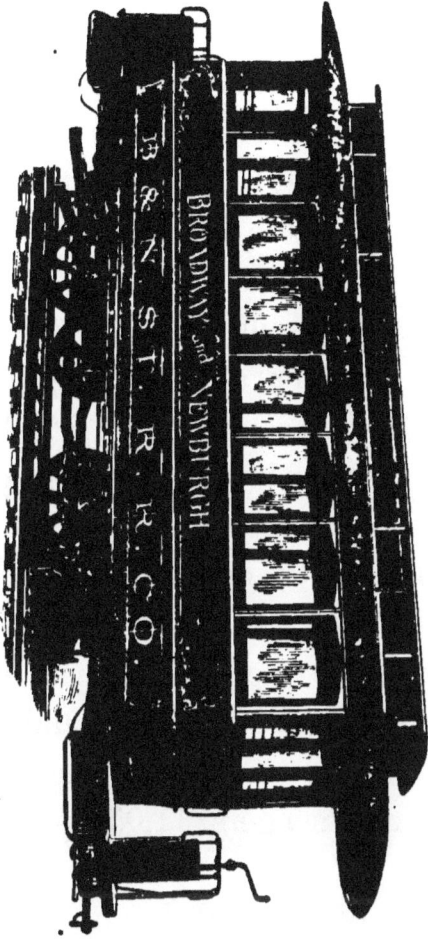

FIG. 34.—SINGLE-TRUCK CAR.

represents a car body 21 feet long and 28 feet in length over all, with a width over wheels of 6 feet, a total width over all of 7 1/2 feet, and capable of seating 30 persons. The truck weighs without motors 3,500 pounds, and the body 5,250 pounds, making a total weight, without motors or passengers, of 8,750 pounds. Fig. 35 shows a double-truck car. The car body is 25 feet long, and 33 feet over all. The width over wheels 6 feet, and over all 7 1/2 feet. This car will seat 36 persons. The weight of the truck without motor is 5,200 pounds, and the body 5,850, making a total weight, without motors or passengers, 11,050 pounds.

A form of *journal box* is shown in Fig. 36. Here the lid *L*, can be moved aside for examination, or for filling the box. The entire box is dust tight. The side frames

FIG. 35.—DOUBLE-TRUCK CAR.

are clamped to and riveted in the grooves
B, B, so that the weight of these frames,
and, therefore, the entire weight of the car

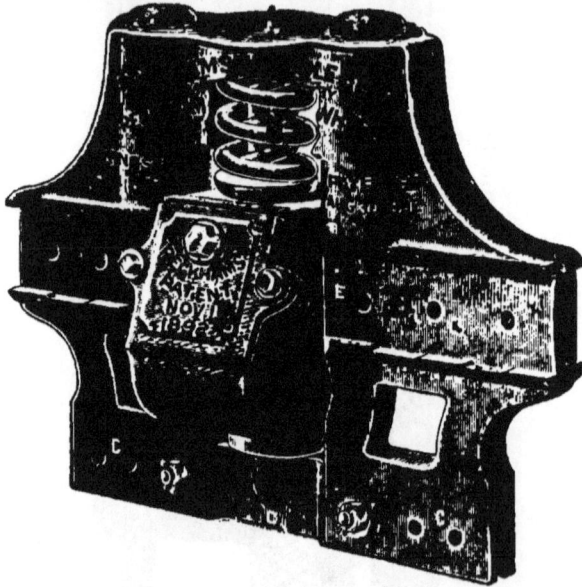

FIG. 36.—FORM OF JOURNAL BOX AND SUPPORT.

pulls down upon the yoke *A*, and presses
the double spiral spring *S*, upon the box
L. The double spiral springs on all the

boxes, therefore, bear the entire weight of the car. Fig. 37, shows a cross-section of

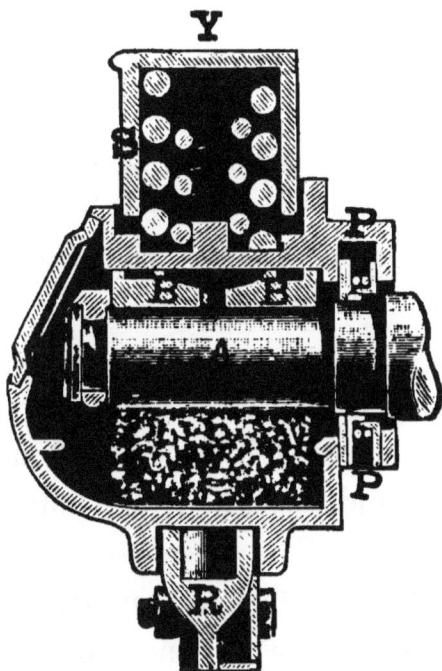

FIG. 37.—SECTION OF BOX SHOWN IN FIG. 36.

these journal boxes taken through the axis of the shaft. *A*, is the axis, *B*, *B*, the brasses from which the weight is transmitted to the axle, W, is the mass of lubricating material, S, the double spiral spring supporting under compression the yoke Y. P, is the spring packing faced with leather to keep out dust. R, is a repair piece which is marked C, in Fig. 36. This repair piece, when removed by withdrawing two bolts, permits the frame to be lifted clear of the axles.

Wheels for electric street cars are usually 30 inches, 33 inches or 36 inches in diameter, and weigh from 300 to 400 pounds each. The *tread* of the wheel; *i. e.*, its running face, is usually chilled to a depth of 1/2 or 3/4 inch to improve its wearing qualities. A good wheel should run 30,000 miles. Wheels are usually forced upon their axles by hydraulic pressure, but in some cases they are bolted to collars on the axle, which collars are themselves forced hydraulically on the axle. There are two types of wheel, the *open* and the *closed*. Fig. 38, shows a form of

FIG. 38.—OPEN CAR WHEELS.

open wheel and Fig. 39, a form of closed wheel.

Motors may be mounted on the trucks in several ways. The most usual method is to support each motor partly on the axle it drives, and partly on a cross beam extending between the side frames. This is shown in Fig. 40, where the motor *M*, of

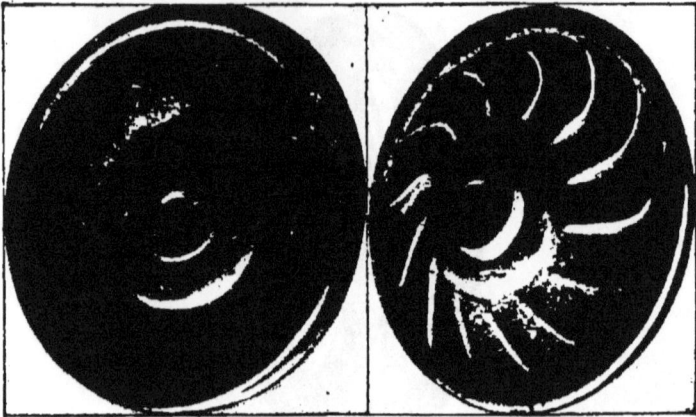

FIG. 39.—CLOSED CAR WHEEL.

the type shown in Fig. 11, is supported on the cross beam *B B*, which is itself supported from the side frames by the spiral springs *s, s.* These spiral springs are, of course, employed to reduce the vibration, or jolting of the motor, when running over an uneven track. The cross beams, instead of passing beneath the motor may pass above it, or on a level with its surface, as shown in Fig. 41, where the beam *B B.*

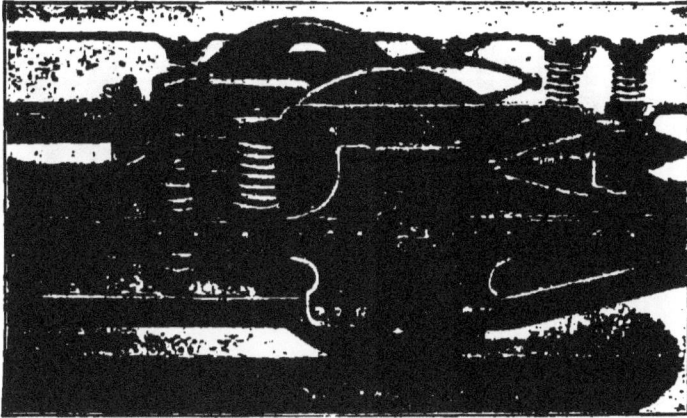

FIG. 40.--METHOD OF MOTOR SUSPENSION.

rests above the spiral springs instead of beneath them. In Fig. 42, another method is shown where the beams *B B*, from which the motor is suspended, are longitudinal and rest on spiral springs, which themselves rest upon cross beams secured to the side frame of the truck. In this case very little of the motor's weight comes immediately upon the driving axle, almost

FIG. 41.—METHOD OF MOTOR SUSPENSION.

all being transmitted to the axle from the side frames.

A plan and side view of the ordinary motor suspension in a single-truck car are shown in Fig. 43, where the two motors *M, M,* are seen, each connected to one of the main axles through the gear *G, G.* The motors are suspended partly upon the main axles and partly upon the cross beams *B B*, and *B B*, the four wheels *W, W, W, W,* are thus directly driven from the motor through the gears.

FIG. 42.—METHOD OF SUSPENDING MOTOR.

The *gearing* employed in connection with the electric street railway cars, is effected by means of a steel pinion upon the armature shaft, such as shown in Fig. 44. This pinion has 14 teeth, which are mechanically cut so as to mesh freely into the teeth of the gear wheel fixed rigidly upon the car axle. This gear wheel is usually made of cast iron in two parts,

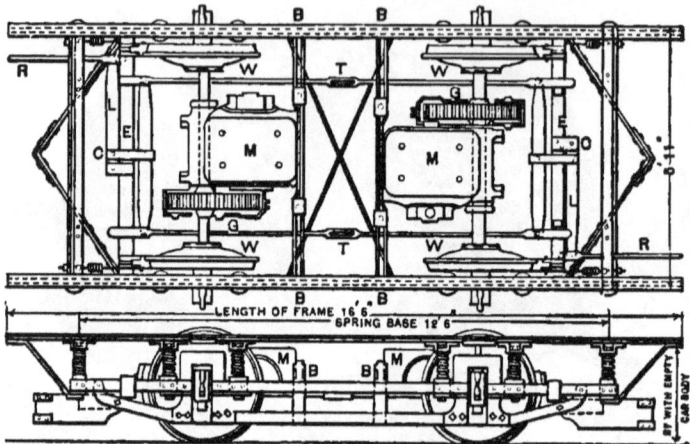

FIG. 43.—PLAN AND SIDE ELEVATION OF MOTOR SUSPENSION.

as shown in Fig. 45. The gear wheel shown has 67 teeth. The ratio of speed reduction between the motor and the car axle is, therefore, in this case, $\dfrac{67}{14} = 4.786$. In other words, the car-wheel axle runs

4.786 times more slowly than the motor shaft. If we consider a car with 33 inch wheels, the circumference of the wheels will be 103.67 inches or 8.639 feet. This will be the distance through which the car will move for one complete revolution

FIG. 44.—ARMATURE PINIONS.

of the wheels. A speed of 1 mile per hour over the track, is a speed of 88 feet per minute, and, therefore, a rotatory speed of $88 \div 8.639 = 10.186$ turns per minute of the car wheels. The speed of the motor armature will be 4.786 times this amount or 48.76 turns per minute. Consequently, for every mile per hour that the car runs, the motors will make 48.76 revolutions per minute. Thus at 10 miles per hour

FIG. 45.—AXLE GEARS.

they will each make 487.6 revolutions per minute.

Pinions are sometimes constructed of hot pressed steel. Thus Fig. 46, shows a steel cylinder before pressing and the completed pinion wheel pressed from such a cylinder.

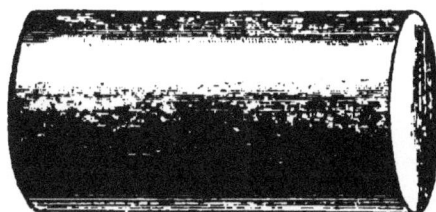

BEFORE.

AFTER.

FIG. 46.—HOT-PRESSED PINION, BEFORE AND AFTER PRESSING.

The motors which we have hitherto considered are all *single-reduction motors*, that is to say, there is only one reduction

in speed effected by gearing between the
motor axle and the car axle. During the
early application of the street car motor it
was very difficult to obtain good *slow-speed
motors* of light weight, and, consequently,

FIG. 47.—DOUBLE-REDUCTION MOTOR.

the expedient was adopted of reducing the
speed down to that required for the car
axle by a *double reduction*. Figs. 47 and
48 show a type of *double-reduction motor*.
In each figure, *A*, is the armature bearing

through which the axle passes. *B*, is an intermediate shaft carrying a *double-gear wheel* at one end as shown in Fig. 48, meshing with a *double pinion* on the armature shaft; while, at the other end, it

FIG. 48.—DOUBLE-REDUCTION MOTOR.

carries a pinion meshing into a gear wheel on the car-wheel shaft passing through the bearing *C.* In this type of machine the double reduction in speed varies from 9 to 19, according to the size of the motor and requirements of speed and power. In recent times the double-reduction motor

has almost disappeared. One difficulty
with the double-reduction motor was the
noise made by the rapidly running armature pinion. To reduce this, *rawhide
pinions;* *i. e.,* pinion wheels made up of

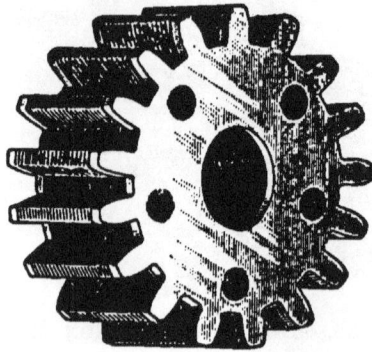

FIG. 49.—RAWHIDE PINION.

discs of rawhide, cut into the proper
shape, assembled and clamped together,
were employed, of the type shown in Fig.
49. The lifetime of such rawhide wheels
was never very extended.

The life of steel and iron gearing depends largely upon the care with which the dust is excluded from them. In prac-

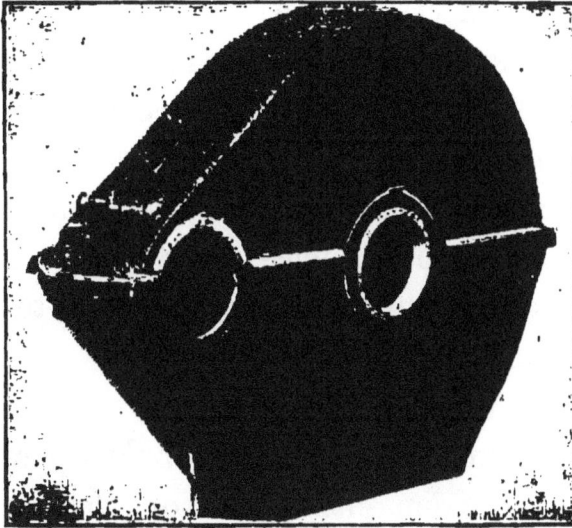

FIG. 50.—GEAR CORES.

tice an increased life is ensured by enclosing the gear in a dust-proof gear cover, as shown in Fig. 50.

It is evident that for safety of running cars through crowded thoroughfares, it is absolutely necessary to be able to stop a car with certainty in a short distance. In order to effect this, various forms of *brake mechanism* are employed. These are either operated by hand, or by the electric current. *Pneumatic car brakes* have not come into any extended use up to the present time for this purpose, since they require the addition of a pneumatic compressor to the car equipment.

A common form of *lever brake*, operated by hand, from either end of the car, is shown in Fig. 51 and also in Fig. 43. *R*, *R'*, are the projecting rods to one or other of which the power is applied by a chain and handle. Fig. 52 shows the ordinary *brake handle* at the car platform. By rotating this handle the chain *C*, is wound upon the handle shaft, thus hauling upon the brake rod R'. P, is a pawl engaging with the pinion wheel on the brake handle shaft so as to hold or release the brake as desired. Fig. 51, shows that when one of

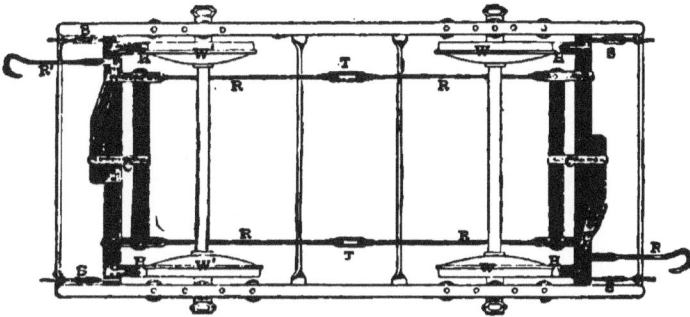

FIG. 51.—HAND BRAKE MECHANISM.

the brake rods, say R, is pulled by the chain, the lever L, is drawn forward and by the action of the short bar C, or brake beam clevis, the brake beam B is forced backwards, so as to cause the brake shoes H, H, to press against the treads of the wheel W, W. At the same the brake frame $L R R R L'$, is forced forward, thus drawing the other brake beam B', forward, and causing the shoes H', H', to bear against the tread of the wheels

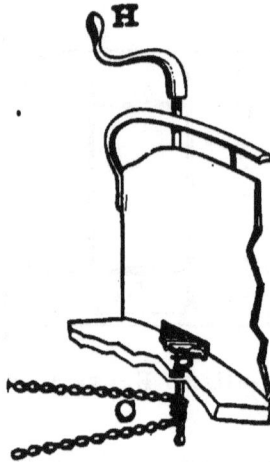

FIG. 52.—BRAKE HANDLE AND CHAIN.

W' W'. As soon as the tension is released from the brake rod, the brake frame $L R R R R L'$, releases and throws the shoes off the wheels.

When the arm is applied to the brake handle *H*, Fig. 52, the pull so delivered is multiplied by the leverage of the handle over the chain. This pull being delivered at *R'*, is again multiplied by the leverage of the brake lever *L*. The combined leverage of the brake staff and brake lever is usually about 50, so that a pull of 100 pounds weight, delivered horizontally at the brake staff handle, represents a pull of about 5,000 pounds delivered at all four brake shoes, or about 1,250 pounds total pressure between each shoe and the wheel it grips. The effect of this pressure is to produce about 1/8th of the pressure as a frictional retarding force, so that if 1,250 pounds pressure be supplied to each wheel, the retarding drag applied at the wheel tread is about 160 pounds.

The turnbuckle *T T*, enables the play

of the brake rods and brake arm to be adjusted so that any unnecessary delay in applying the brakes may be avoided.

A form of *electric car brake*, which

FIG. 53.—ELECTRIC BRAKES MOUNTED ON STREET CAR TRUCKS.

promises to come into extended use, is represented in Fig. 53. A truck is here represented with two motors *M*, *M*, in place, of the same character as shown in Fig. 11. In addition to the ordinary hand brake mechanism operating through the brake rods *r*, *r*, the brake levers *l*, *l'* the brake beam *m*, and the shoes *b*, *b*, there is supplied an electric brake *B*, on each car axle. This brake is in two parts; namely, a cast iron disc *C*, rigidly keyed to the car wheel axle, and, therefore, revolving with the car wheel, and a circular shoe or compact electromagnet *D*, facing *C*, clamped to the motor and frame of the car, and, consequently, not rotating whether the car be running or at rest. When the car is running there is no friction between the shoe *D*, and the disc *C*. As soon as it is desired to stop the car, the trolley circuit is first broken at the trolley switch by the motorman, thus cutting off the power from the line. As soon as this is done the motors which are still running by the momentum of the car, act as ordinary dynamos, and are capable of furnishing a temporary electric current

as soon as a circuit is closed to their
E. M. F. The coil of insulated wire in the
interior of the magnet shoes D, D, of the
brakes are placed in circuit with the motor
armatures so as to receive this current.

Under these circumstances a powerful
electromagnetic attraction occurs between
the shoes D, D, and their iron discs C, C,
tending to clutch them together and stop
the wheels. The faster the car is running
at the moment these brakes are applied,
the more powerful is the current that is
generated by the motors acting as dynamos,
and, consequently, the higher the brake
action.

There are two methods of controlling
this brake, the first automatic, and the second under the control of the motorman.
The braking power, if uncontrolled, would

be so great that the wheels would be instantly locked and would *skid* or slide on the track. An *automatic switch* is placed in the circuit in such a manner that the current strength from the motors through the brakes is limited to that which will apply the maximum braking power without permitting skidding with a light car. Moreover, the braking current passes through the controller to be subsequently described, and is thus regulated in strength by the motorman, so that he can apply the electric brake either suddenly, or gradually, as he may desire. The advantage of the electric car brake is the power it possesses, the swiftness with which it can be applied, and the fact that it is independent of all current taken from the trolley wire, since the moving motors supply the energy needed. The mechanism can, moreover, be attached to any car without great expense, while

the ordinary brake is left untouched for
use in cases of emergency. A special ar-
rangement is made to lubricate the rotating
surfaces by means of a graphite brush carried in the shoes D, D. This prevents excessive wear and heating; for, in this brake,
the retardation is very largely a magnetic
pull rather than a mechanical friction, and,
in this way, effective brake action is
secured without excessive rubbing.

When the rails are slippery, by reason of
a thin film of mud or frost, an application
of the brake is apt to cause adhesion of the
shoe to the brake wheel, and a skidding or
slipping of the wheel on the track, instead
of an adhesion between the wheel and the
track and a slipping of the brake shoe on
the wheel. The result of this skidding is
to wear the tread of the wheel at the point
of its periphery at which it slips along the

track, whereas in the normal application of the brake, this wear is uniformly distributed over the entire wheel surface against the brake shoe. Under these conditions the wheel tends to flatten at the point of skidding, and once a depression is formed, there is a continual tendency to increase the amount of flattening. Flat wheels are not only difficult to brake properly, but produce an uneven jarring motion very disagreeable to the passengers. In order to increase the adhesion between the wheel and track, so as to be greater than that between the brake shoe and the wheel, sand is sometimes poured upon the track with the effect of producing a greater friction. Various forms of *sand boxes* have been devised for sprinkling a small quantity of sand directly beneath the wheel on the track where it is required. One of these forms is shown in Fig. 54. The sand box

S, is mounted within the car close to the platform. The motorman, by pressing with his foot upon the foot-button *F,* depresses the lever *L,* which is pivoted at *P,* and

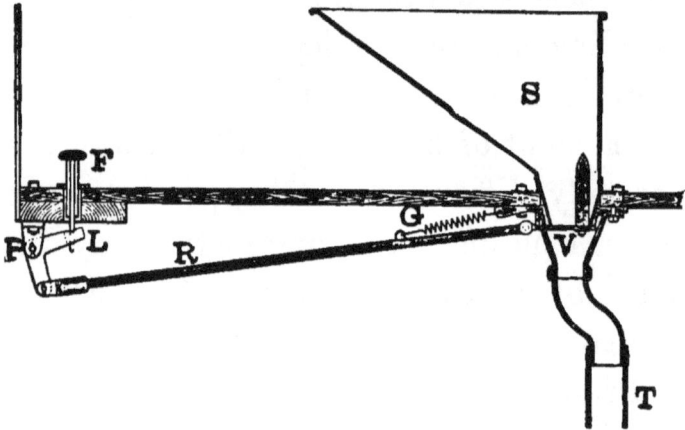

Fig. 54.—Sand Box.

thus causes the rod *R,* to move forward in the direction of its length against the tension of the spiral spring *G.* This opens the valve outlet and allows sand to pour through the tube *T,* upon the track beneath.

On the truck of a car there is mounted a car body familiar to all our readers. These bodies are of four types; namely, the open or summer car, the closed car, the convertible car, and the double decker. The latter is not in use on overhead trolley lines.

CHAPTER VI.

ELECTRIC LIGHTING AND HEATING OF CARS.

THE advantages possessed by electric lighting, as obtained from incandescent lamps, are so evident, that this method of artificial illumination is almost invariably employed in trolley cars. The current required for the lighting of these lamps is, of course, taken from the same source which drives the car, that is to say, a special circuit is taken from the trolley to the track, through the lamps to be lighted. The type of incandescent lamp employed varies with the number placed in the car. If, as is commonly the case, there are five lamps, three in the centre and one at each end, they are connected in series, so that the current passes successively through each, and they are placed in a special cir- cuit directly between the trolley and the track as represented in Fig. 55. Here

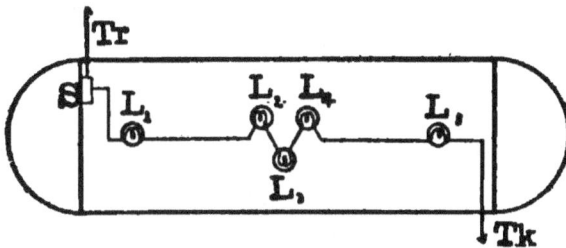

FIG. 55.—DIAGRAM OF LAMP CIRCUIT OF CAR.

the wire leading to the trolley wheel is marked Tr, and enters the switch S, from which it passes through the five lamps L_1, L_2, L_3, L_4, L_5, in succession, finally pass- ing to the track Tk, through the frame- work of the truck.

In this case since the total pressure be- tween the trolley and track is approxi-

mately 500 volts, and there are five
lamps in series, the drop in each lamp will
be 100 volts, the current strength being
about 2/3rds ampere. The total activity
developed in the lamps will be roughly
500 volts × 2/3rds ampere = 333 watts,
or less than one-half of a horse-power.
When nine lamps are employed to light
the car, in three clusters of three each, all
nine are placed in one series, the drop in
each lamp being approximately $\frac{50}{9} = 55.5$
volts. The current strength in this case
will be a little more than 1 ampere and
the activity in the lighting circuit will
be nearly 500 volts × 1.1 amperes = 550
watts, or 3/4ths horse-power. This activity
has to be sustained during the operation
of the cars at night time whether the car
be running or not. If five lamps are
employed, each lamp must be made for a

pressure of roughly 100 volts, while if nine lamps are employed, each lamp must be made for a pressure of roughly 55 volts.

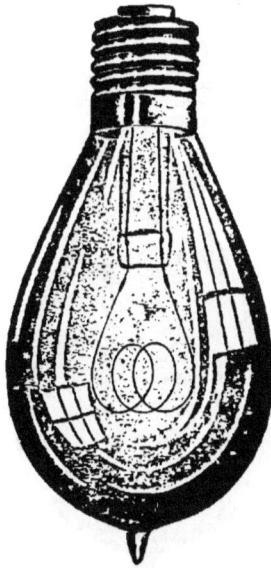

FIG. 56.—CAR LAMP.

Fig. 56 shows a common form of lamp employed in street cars. Fig. 57 shows another form in which the incandescing filament is anchored or supported at its

centre for the purpose of preventing the
filament from being injured by excessive
vibration. Incandescent lamps for street

FIG. 57.—RAILWAY LAMP WITH ANCHORED OR NON-
VIBRATING FILAMENT.

car use have usually an *efficiency* of 1/4th
candle per watt; *i. e.*, when operated at
the pressure for which it is designed, it

gives normally 1/4th of a candle per watt of activity absorbed, so that a 16 candle-power lamp would require normally 64 watts.

FIG. 58.—FORM OF FIXTURE FOR CAR LAMP.

A common form of lamp fixture is shown in Fig. 58 and a cluster suitable for three lamps is shown in Fig. 59.

A form of switch for turning the car lamps on and off, is shown in Fig. 60. This switch box is screwed up inside the car near the ceiling and has a projecting key K, for turning the lamps on or off. The action of the key is illustrated by the switch shown in Fig. 61, where A and B,

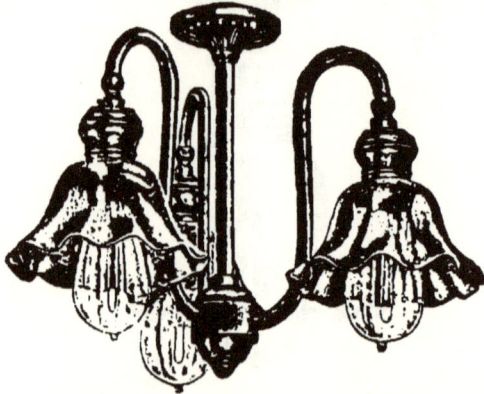

FIG. 59.—FORM OF THREE-LAMP CLUSTER FOR CAR.

are the binding posts connected to one side with the trolley wheel and the other with the lamps. On turning the key D, the brass piece C, may be made to bridge metallically across between the posts A and B, thus closing the circuit through all the lamps. The switch box, in Fig. 60, also contains a *safety fuse* or *cut-out.* This simple device consists of a wire of lead or other alloy that will melt, and thus automatically break the circuit, if the current becomes excessive.

Fig. 60.—Switch and Cut-out for Car Lamps.

It will be evident that for every 16 candle-power incandescent lamp operated in the car, about 64 watts activity will be required; or, roughly, 1/12th of a horse-power per lamp at the car, which may represent say 1/8th of an indicated horse-power at the engine.

When street cars are running in cold climates the artificial heat required at certain seasons of the year may be obtained either by the use of an ordinary coal stove,

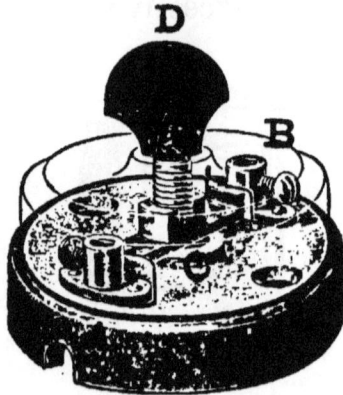

FIG. 61.—SWITCH FOR CAR LAMPS.

or by electric heaters. Although the coal stove is the cheaper of the two, yet it possesses several inconvenient features. In the first place it occupies useful space; in the second place it requires attention and introduces more or less dust, smoke or dirt into the car, while the heat which it gives is principally developed in the upper portion of the car, the air near the floor remaining comparatively cold. Moreover, some time is required to start a fire in a stove.

In contrast with these inconveniences, the electric heater possesses such marked advantages, that, despite its extra cost, it has come into use for the heating of electrically propelled cars. When an electric current passes through a wire, heat is developed therein. Thus, we have already seen that when a current passes through a trolley wire, a certain amount of power will be expended in heating the trolley wire. Under practical conditions the trolley wire will never get sensibly warm by the current it carries for the reason that the surface it freely exposes to the air is so great, that, taken in connection with its mass, the comparatively small amount of heat developed within it is rapidly liberated. If, however, the same amount of electric resistance which exists in a mile of trolley wire, were obtained

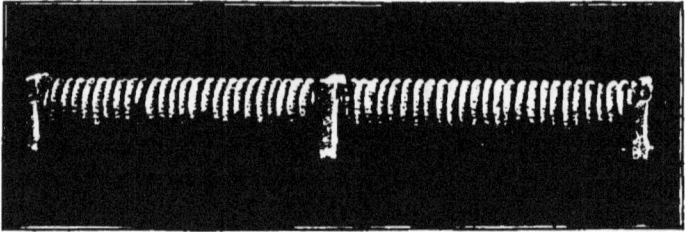

FIG. 62.—HEATING COILS OF CAR HEATER.

in a short length of copper or iron wire. then the same amount of heat would be produced in a much smaller mass of iron, having a greatly reduced surface, with the result of producing a much higher temperature in the wire.

The coils of wire used in a particular form of car heater are shown in Fig. 62.

Here the heating coil consists of galvanized iron wire which is wrapped in the form of a close spiral and then placed in a spiral groove on the outside of a porcelain tube. This construction affords a great length of heating coil in a small space, so supported as to prevent the coil changing its form when heated and yet practically permitting nearly all of its surface to give off heat to the surrounding air. In the heating coil shown in the figure, which is about 3′ 6″ long, there are 392′ of wire; the size of wire being No. 20 A. W. G. iron wire, having a diameter of 0.032″, or 32 mils. The total surface exposed by the coil in a single heater is 1.642 square feet. The coil is placed in a metal case, so provided with openings as to permit the free flow of air entering at the bottom of the case to flow around the heater, come in contact with the heated wire and to

escape through a grating at the top.
When so desired, the air may be taken in
directly from the outside of the car. The
coil in its metal case, ready for fastening
in position below a seat, is represented in

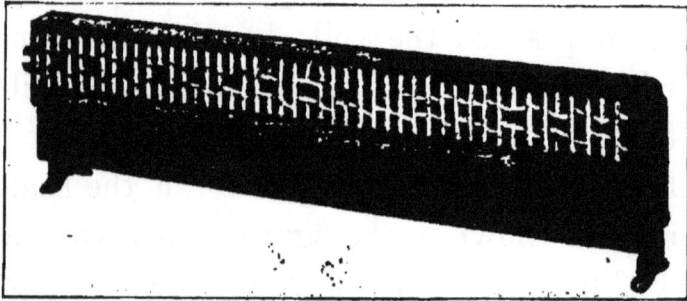

FIG. 63.—ELECTRIC CAR HEATER.

Fig. 63. The heater is sometimes placed
with its grating flush with the riser beneath the seat. In this case the form of
heater is that shown in Fig. 64. For cars
of the ordinary size, four or six heaters are
employed; *i. e.*, two or three on each side
of the car. The heaters are placed in the

risers of the seats near the floor. In Fig. 65 the interior of a car is shown equipped with six heaters, four of which are seen beneath the seats at the points A, B, C, D.

FIG. 64.—FORM OF ELECTRIC CAR HEATER.

In order to regulate the amount of heat required to meet the changes in temperature, a *temperature-regulating switch* is employed, by means of which the separate heaters may be connected in series or in parallel groups between the trolley and the track, or by means of which one or more of the heaters may be removed at will. By this means the amount of cur-

rent which passes through the heaters, and, therefore, the amount of heat they develop can be adjusted. When the switch is turned, so as to place all the heaters in series, the resistance in the heating circuit is greatest and the heat produced is least. When all the heaters are employed in three parallel groups of two each, the maximum current is supplied and the maximum heat is obtained.

Fig. 66, represents the interior of the temperature-regulating switch, by which these varied connections are made. Fig. 67, shows the exterior appearance of the switch. There are five positions of this switch when the current is passing through it, numbered respectively, 1, 2, 3, 4 and 5, and the particular position is indicated by the numeral appearing through the opening at *W*, in the switch casing. The

Fig. 65.—Interior of a Car Equipped with Electric Heaters.

switch is so constructed that before changing from one number to another, the circuit of the heaters is opened. In position
5, as shown in the figure, the full current

FIG. 66.—INTERIOR TEMPERATURE-REGULATING SWITCH
(FIVE INTENSITIES).

strength of about 12 amperes passes
through the heater, representing a total
activity of about 500 volts × 12 amperes =
6,000 watts = 6 KW = 8 HP approximately. This activity is entirely expended
in heating the wire, and, therefore, in

warming the air which comes in contact with the wire. Position No. 1, corre-sponds to the minimum activity and allows about 2 amperes to pass through the

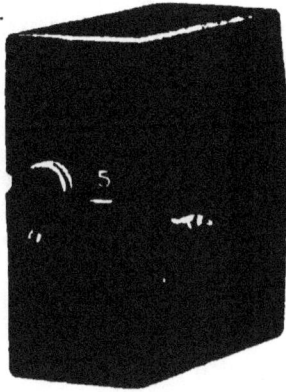

FIG. 67.—EXTERIOR TEMPERATURE-REGULATING SWITCH.

heater, representing a total activity of about 500 volts × 2 amperes = 1,000 watts = 1 KW = 1 1/3 HP, approximately. In practice, it is found that in cold weather about 6 amperes have to be main-tained in the heaters, representing an activity of roughly 3 KW. The cost of producing a KW-*hour*, or 1,000 joules-per-second for 3,600 seconds = 3,600,000 joules,

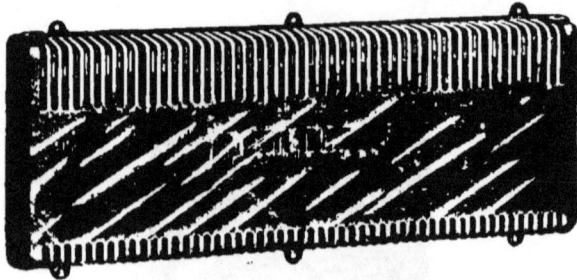

FIG. 68.—CAR HEATER.

varies considerably with the size of the electric plant supplying the current, but, speaking generally, a fair average may be considered as being 1 1/2 cents per KW-hour, so that the expense of heating the cars electrically during severe weather may be estimated roughly as 4 1/2 cents per hour.

Another form of car heater and its enclosing case is shown in Figs. 68 and 69.

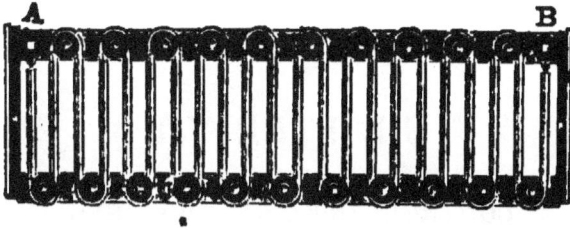

FIG. 69.—CAR HEATER, DESIGNED TO ATTACH TO SEAT RISER.

This operates on practically the same principles.

CHAPTER VII.

It is necessary in the practical operation of a street car to place both its speed and the direction of its running under the control of the motorman. Moreover, the apparatus employed to do this should require for its operation no more than ordinary intelligence, that is, should be capable of being operated without any electrical skill on the part of the motorman. On electric trolley cars, as is well known, the motorman controls the car by means of two handles, the right hand one of which controls the mechanical brake apparatus and the left hand one the electric apparatus called the

controller. This latter apparatus is contained within a vertical metal case
provided on its upper plate with notches,
corresponding to different speeds of the
car. By this apparatus the electric current is turned on and off and the power
and speed of the motor controlled.

Different systems of electric traction
employ different forms of controllers, but
all operate on essentially the same plan.
It will, therefore, suffice, in pointing out
the method in which the controller
operates, to limit the description to a particular form in common use.

The external appearance presented by
this controller will be seen by an inspection of Fig. 70. One of these controllers
is mounted on the front platform and a

similar one on the back platform of the car. It is operated by the movement of the handle *H*. The small handle *h*, controls an *emergency switch* used for reversing the motion of the car when necessary. Coming now to the controller, *S*, is a stop to limit the range of motion of the handle. In the position shown the current is turned off, and, as the handle is turned around in a clockwise direction, the motors are gradually brought into action with increasing speed, until, when the projection of the handle strikes the stop *S*, on the other side, after nearly one revolution, the maximum speed of the car is attained.

In order to open the controller, a sheet iron door is provided, closed with screw bolts, which can be manipulated by hand, the hinges of these bolts being shown at *j, j.*

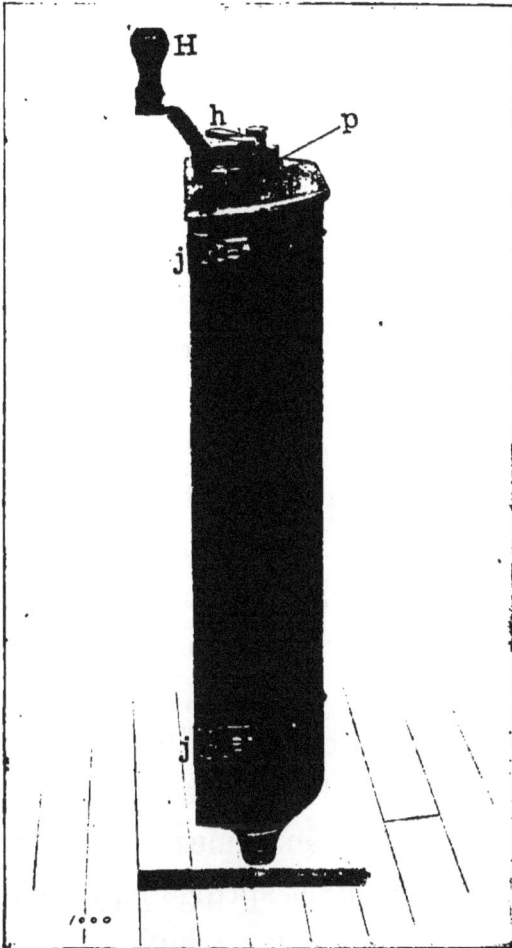

FIG. 70.—CONTROLLER, CLOSED.

The interior construction of the controller is shown in Fig. 71. The lid $L\,L$, has been thrown back by withdrawing the bolts at the hinges j, j. By further withdrawing the small bolt, J, shown separately beneath the lid, an iron cover C, hinged on the core c, of the electromagnet M, is also thrown back from the cylinder Y, leaving it exposed to view.

The switch cylinder is turned by the movement of the handle H. It carries eleven rings of insulating material r_1, r_2, r_3, etc., upon which are mounted metallic conducting segments s_1, s_2, s_3, of different lengths and in different positions, so that, when the cylinder is turned, they come into contact at different times with the row of eleven fixed contact springs p_1, p_2, p_3, p_4, etc. It is these contacts which effect the changes in the connections for producing a

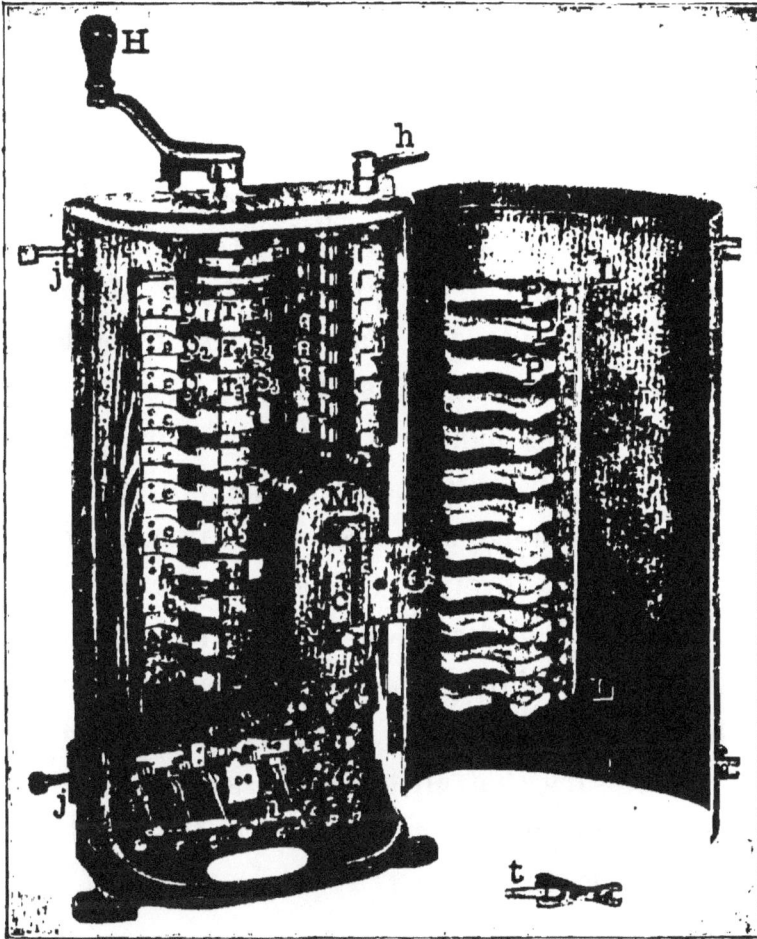

FIG. 71.—CONTROLLER, OPEN.

change in speed of the car. In the position shown, while the handle is at the first notch and against its stop, none of the segments are in contact with their springs, and the trolley is disconnected from the motors. The small handle h, rotates a small cylinder y, carrying eight sets of metallic segments and having a row of eight fixed metallic contact springs q_1, q_2, q_3. By throwing the handle h, over about 60°, the segments in contact with the springs q, can be changed and also the direction of the current through the armatures of the motors. The direction of rotation of the motor can thus be reversed, backing the car. At W, is a star wheel, which renders it necessary that all the successive contacts be made and none omitted when the handle is turned.

After contact has been made between

the motors and the trolley, so as to pass the usual current strength through the circuit; then, on breaking contact either at the trolley, or at the ends of two wires in the circuit on the car, a spark or metallic arc will form, which may be from two inches to five inches in length. This is the characteristic arc which is seen when the trolley wheel jumps from the wire. It will be readily understood that the formation of arcs of this character within the controller, would soon cause its destruction. This is avoided in the form of controller shown in the figure by means of a device called the *magnetic blow-out.* The current through the motors passes through a coil wound on the magnet M, around the iron core c. This makes the core c, a powerful electromagnet, and its projection, or pole-piece C, becomes a large magnetic pole. When this pole-piece is close down in its normal position the polar ridges P, P, P, rest close to the contact strips p_1, p_2, p_3. While the motors are running, the magnet M, being excited, produces a powerful magnetic flux surrounding the contact surfaces p_1, p_2, p_3. As soon as any break in the circuit occurs either in changing connections during running of the motors, or particularly when the current is entirely shut off, the severe sparking, which would occur, is prevented because the arcs are blown out under the influence of the magnetic flux from this magnet. In other words, an arc cannot be maintained in the presence of a sufficiently powerful magnetic field.

It remains now to explain the manner in which the different positions of the handle H, alter the speed of the car. There are altogether eleven notches or successive positions which the handle H, and, consequently, the switch cylinder, can assume. The first corresponds as already mentioned, to no current, as in Fig. 71. There are thus left ten positions, at which the speed of the motor can be varied.

When the handle is pushed to the first working position, the segments s_1, s_2, s_3, engage with their corresponding springs. The effect of this is to establish the connections shown in Fig. 72. R_1, R_2, is a resistance made up of a coiled insulated strip of iron. M_1 and M_2, are the motors. It is evident, therefore, that the current has in this case to pass successively from the trolley T, through the resistance R_1, R_2, and the two motors to the ground G. The resistance R_1, R_2, may be 1 ohm, that of each motor armature 0.4 ohm, and that of each field 0.8 ohm. The total resistance of the circuit between the trolley and the track is, therefore, 3.4 ohms.

If we assume that the usual pressure is steadily maintained between trolley and

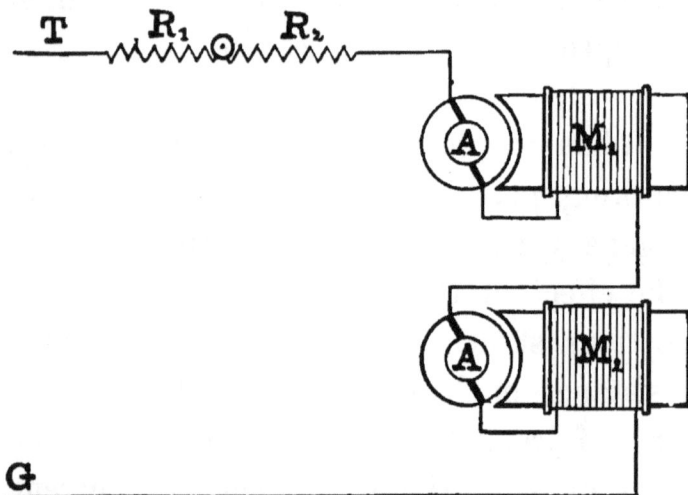

FIG. 72.—CONNECTIONS CORRESPONDING TO FIRST WORK-
ING NOTCH OF CONTROLLER.

track at 500 volts, then the maximum current strength which may pass through the car circuit under these conditions is, by Ohm's law, 500 volts divided by 3.4

ohms = 147 amperes. In practice the current never rises to this amount for two reasons; namely,

(1) As soon as the circuit is closed, the excitation of the field magnets causes a powerful development of magnetic flux in the motors, which momentarily sets up a C. E. M. F. tending to check or oppose the establishment of the current. This C. E. M. F. is of very brief duration, say about one second, so that if the motor was prevented from running, the full current strength according to Ohm's law would soon be reached. This is called the *C. E. M. F. of self-induction,* because it is produced by the magnetic inductive effect of the current on its own circuit.

(2) As soon as current passes through the motor it begins to turn and in so doing acts as a dynamo to produce a C. E. M. F. which permanently checks the

current, and the faster the motor runs the greater this C. E. M. F. This *C. E. M. F. of rotation* is far more important than the C. E. M. F. developed by self-induction, since it always operates while the motors are running, whereas the C. E. M. F. of self-induction only exists during changes in current strength.

It now remains to be explained how the C. E. M. F. of rotation automatically regulates the strength of current and, therefore, the amount of electric activity supplied to the car. Let us first suppose that the motors are disconnected from the car axles, and allowed to revolve freely without any friction whatever. If such a state of things were possible, the torque, or rotary effort of the armature produced by the current which first enters them, would soon bring the armatures to a high rate of speed, under which circumstances the C. E. M. F. generated by them would be so great that very little current would pass through them. Thus, if the total resistance of the motor circuit, as shown in Fig. 72, was 3.4 ohms, and the amount of power required to drive the motors light and frictionless, was only 1 HP, or say 750 watts, this would mean a current strength of but 1.5 amperes, since 500 volts × 1.5 amperes = 750 watts. In order to limit the current strength to 1.5 amperes in a circuit of 3.4 ohms, the effective pressure must be 5.1 volts, since $\frac{5.1 \text{ volts}}{3.4 \text{ ohms}} = 1.5$ amperes. The effective pressure between trolley and track must, therefore, under these circumstances, be only about 5 volts, and this will be produced by such a speed of the motor as to develop a C. E. M. F. of 495 volts; for, since 500 volts is the E. M. F. supplied, and 495 is the C. E. M. F., the difference serving to drive the necessary 1.5 amperes through the resistance of the circuit will be 5 volts.

If now some small frictional resistance or load be applied to the motors; or, in other words, if the motors be required to do some little work, the activity which they will require to be supplied with to perform this work may amount to 2 HP or, approximately, 1,500 watts (1.5 KW). Under these circumstances the current strength must increase to 3 amperes, since 3 amperes at a pressure of 500 volts represents the needed activity of 1,500 watts. In order to permit 3 amperes to pass through the resistance of the circuit (3.4 ohms) the effective pressure must be 10.2 volts, since 10.2 volts ÷ 3.4

ohms = 3 amperes. The E. M. F. applied being 500 volts, the C. E. M. F. must be 490 volts and the speed of the motors will drop sufficiently to produce only 490 volts C. E. M. F., instead of 495. In the same way if we suppose the motors to be coupled to their respective car axles, and work to be required from them to drive the car, to an amount of say 10 HP, then the power which must be supplied to the motors to make up for losses, both frictional losses in the gears and bearings, and electrical losses in the armature and field coils, may be 15 HP, or 15 × 746 = 11,190 watts = 11.19 KW. This is the electric activity which must be supplied from the circuit to the motors, and will represent a current strength of 22.38 amperes at a pressure of 500 volts. The effective pressure required to drive 22.38 amperes through a resistance of 3.4

ohms, will be 76.09 volts. The C. E. M. F. required to limit the effective pressure to approximately 76 volts, will be 500 − 76 = 424 volts, and the motors will drop in speed until this is the C. E. M. F. which they supply.

Proceeding in this way, the more *load* we put on the motors; *i. e.*, the more we load the car, or the steeper the grade it is necessary to ascend, the greater the electric activity which must be supplied to drive the car, and the greater the current strength which must be passed through the motors to produce this activity. Under these circumstances the motors will continue to slacken in speed so as to permit the current to pass, and will always attain such a speed as will permit the required activity to enter them in order to perform the work they have to do. When, finally, the load is so great that the motors are unable to run, the activity received will be that defined by Ohm's law, shortly after the circuit is closed. In this limiting case, all the activity is expended as heat in the resistance R_1, R_2, and in the motors M, M. In all other cases, when the motors are running, some of the activity is developed as heat, but by far the greater part is developed as mechanical activity in propelling the car.

On the other hand a reduction in the load of the motors must be followed by an increase in their speed. This increase, however, will be arrested as soon as the C. E. M. F. is increased to the value which limits the current strength to that required for the rate at which work is being mechanically done.

In all cases it will be evident that the E.

M. F. existing between the trolley and the track, which we have assumed to be maintained at 500 volts, is to be met by an equal total C. E. M. F. in the car circuit *T, G.* If the motors are at rest, this C. E. M. F. must be entirely due to drop in the resistance, represented by the product of the current strength in amperes and the resistance in ohms. For example, with the car held at rest, we know that the current will be 147 amperes in the case of Fig. 72, and this multiplied by the total resistance of 3.4 ohms, represents a drop of pressure amounting to 500 volts.

When, however, the motors are running, their C. E. M. F. of rotation will necessitate a smaller drop in the resistance of their circuit. Thus, if the motors are producing together a C. E. M. F. of 400 volts, then the drop in the resistance R_1, R_2, will

be only 100 volts, and if the motors produced together a C. E. M. F. of rotation of
490 volts, the drop will be reduced to 10
volts. The activity available for mechanical work is the product of the C. E. M. F.
of rotation and the current strength. For
example, if the motors in Fig. 72 develop together a C. E. M. F. of rotation amounting
to 490 volts, and the drop in the resistance
of motors and rheostat is, therefore, 10
volts, the current strength, which will produce this drop, will be by Ohm's law
$\dfrac{10 \text{ volts}}{3.4 \text{ ohms}} = 2.94$ amperes, approximately.
The total activity taken from the circuit
between T and G, will, therefore, be 500
volts × 2.94 amperes = 1,470 watts. Of
this, the amount capable of producing mechanical activity is 490 volts × 2.94 amperes
= 1,440 watts, while that only capable of
producing heat is 10 volts × 2.94 amperes

= 29.4 watts. It is evident that the greater the proportion of C. E. M. F. of rotation to the drop in the circuit T, G, for any given current the greater will be the activity used for propelling the car.

We will now explain the use of the resistance R_1, R_2. In the first place if the resistance R_1, R_2, be removed from the circuit in Fig. 72, the total resistance between T, and G, will be reduced to 2.4 ohms, and the possible current strength by Ohm's law, such as would exist when the car was absolutely prevented from moving, would be $\dfrac{500 \text{ volts}}{2.4 \text{ ohms}} = 208$ amperes, approximately. In other words, a current of 208 amperes would maintain a drop of 500 volts in a total resistance of 2.4 ohms. The first rush of current would, therefore, be greater, and the current strength dur-

ing the time when the motors were accelerating and reaching their limiting speed of
rotation would be greater, so that the car
would start from rest with a greater jerk,
and, moreover, waste a greater amount of
power in the process. The greater the
amount of resistance which is introduced
into the circuit of the motors at the start,
the smaller the current which will pass
through them, the more quietly the car will
start and reach the speed which limits the
C. E. M. F. of rotation, and the less the
activity which will be wasted during that
period in which the motors are accelerating
up to this speed.

On the other hand, the continued use of
a resistance R_1, R_2 is more or less wasteful
after the car has been brought up to speed,
because it produces a drop in the circuit
and prevents the C. E. M. F. of rotation

from coming into full play. For example, if the current which the circuit must receive under given conditions of load is 50 amperes, the drop in the resistance of 1 ohm at R_1, R_2, will be 50 amperes × 1 ohm = 50 volts, and the effect is temporarily the same as though the motors M_1, M_2, were connected to the trolley circuit between T and G, without a resistance, but with 450 volts pressure. The activity expended in the motors, both as drop in their resistance, and as available energy against their C. E. M. F. of rotation, will be 450 volts × 50 amperes = 22,500 watts. The circuit between T and G, supplies a total activity of 500 volts × 50 amperes = 25,000 watts.

The effect, therefore, of constantly maintaining the resistance R_1, R_2, in the circuit of Fig. 72, is to expend activity in it as heat, and thus prevent the motors from reaching as high a speed as they otherwise would, while, of course, it is an advantage to be able to run slowly, it is nevertheless

FIG. 73.—STREET CAR RESISTANCE COIL.

a disadvantage to waste power in the re-sistance for this purpose. The use of a certain amount of resistance is, therefore, beneficial during periods of starting, and where the advantage of running at low speeds offsets the disadvantage of wasting power.

The resistance R_1, R_2, is commonly made in the form shown in Fig. 73. Here the coils are placed in an iron box of such dimensions as to permit it to be attached by screws or bolts at the lower part of the car body. It is purposely left open to permit the circulation of air and thus carry off the heat generated in the coils.

Let us now inquire what happens when the controller is turned to the next or second working notch. The effect of this is shown diagrammatically in Fig. 74. An inspection of the figure will show that half the extra resistance is cut out of circuit; namely, R_1. This has the effect of reducing the drop of pressure in the resistance for a given current strength passing through the circuit. Consequently, the motors have to run faster to make up the total C. E. M. F. of 500 volts, so that the speed of the car is

increased.   For example, if in the case of
Fig. 72, a drop of 100 volts occurs in the
resistance R_1, R_2, requiring 400 volts to be

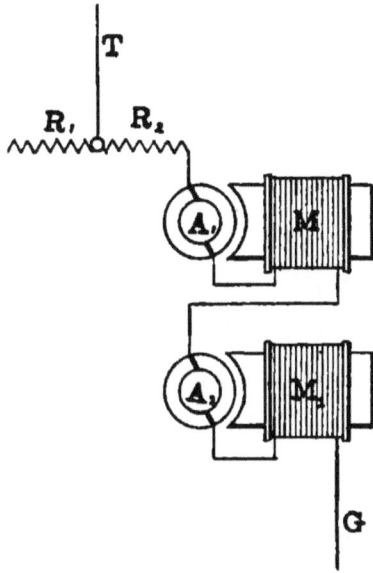

FIG. 74.—CONNECTIONS CORRESPONDING TO SECOND
WORKING NOTCH OF CONTROLLER.

made up by the motors in C. E. M. F. of
rotation and drop in their resistances, then,
when the resistance R_1, is cut out, as in Fig.

74, with the same current strength there will be only 50 volts drop in that half of resistance R_2, remaining in the circuit, and 450 volts must be made up by the two motors in C. E. M. F. of rotation and drop; they will, therefore, increase in speed to this extent. Consequently, the effect of cutting out resistance from the circuit is to cause the car to increase in speed to an extent which will depend entirely upon the amount of drop reduced, which in its turn will depend upon the load of the car. If the car is very light, and is steadily running on a level portion of the track, the drop in the resistance will be very small and the effect of halving this drop will be very small, so that the car will receive very little increase in its steady speed by moving to the second notch. If, on the contrary, the motor is heavily loaded, or is running up a steep grade, there will be a heavy

drop in the resistance R_1, R_2, especially on
starting, due to the stronger current and
greater activity required, so that cutting
out half the resistance and drop will produce a greater increase in speed.

Fig. 75, shows the effect of turning the
controller handle to the third notch. Here,
as will be seen, all the resistance R_1, R_2, is
cut out, so that the motors have to make
up in drop and C. E. M. F. of rotation, the
full line E. M. F. existing between trolley
and ground. They will, therefore, require,
other things remaining the same, to maintain a higher speed than in either of the
preceding positions of Figs. 72 and 74.
The total resistance of their circuit between
T and G, is 2.4 ohms.

Turning the controller handle to the
next, or fourth working notch, the effect

produced is represented diagrammatically
in Fig. 76. Here, as in Fig. 75, the extra
resistance is entirely cut out and in addition

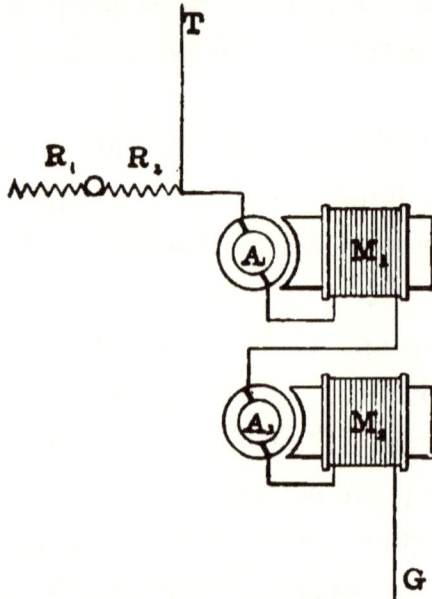

FIG. 75.—CONNECTIONS CORRESPONDING TO THIRD
WORKING NOTCH OF CONTROLLER.

the field magnet coils of each motor are
provided with a *by-path* or *shunt* S_1, S_2, so
that the current through the circuit divides

at each field magnet, a part only going through the magnets, and the remainder going around through the shunt, all of the

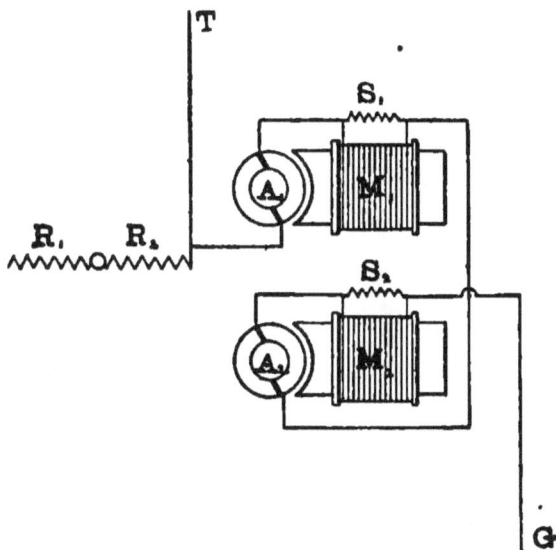

FIG. 76.—CONNECTIONS CORRESPONDING TO FOURTH
WORKING NOTCH OF CONTROLLER.

current, however, passing through each armature.

The effect of shunting the field magnetizing coils is to weaken them; i. e., has

the same effect as taking wire off the coil,
or of reducing the equivalent current
strength. The magnetism produced by
the field magnets of the motors, will,
therefore, be reduced, and in order to
make up a given C. E. M. F., with this
reduced magnetism, a greater speed must
be attained by the armatures. The car
has, therefore, to run faster owing to the
introduction of the shunts. At the same
time, if the grade and load remain the
same the greater speed of the car will call
for a greater expenditure of mechanical
power, and, consequently, a greater expenditure of electric current and activity,
so that, since each motor is called upon to
produce a total C. E. M. F. of 250 volts
in drop and in rotation, this C. E. M. F.
will be developed by a greater speed in
the weakened magnetic fields, but with a
greater current supply and to that extent

a greater drop; for, if the current strength supplied was insufficient to maintain the increased activity of the car, then a decrease in speed would occur until the current supply was made up.

The connections corresponding to the fifth working notch are the same as those shown in Fig. 74, that is to say, the resistance R_2, is first restored to the circuit before changing the connections at the next step.

The condition of affairs when the controller is turned to the sixth working notch is represented in Fig. 77. Here, it will be observed that the shunts around the field magnets are withdrawn, and the resistance R_2, is restored to the circuit, while the second motor M_2, is completely cut out. The first motor M_1, has now to

make up the full pressure of the line with
the aid of the drop in half the resistance.
Excluding the drop in the resistance R_2,

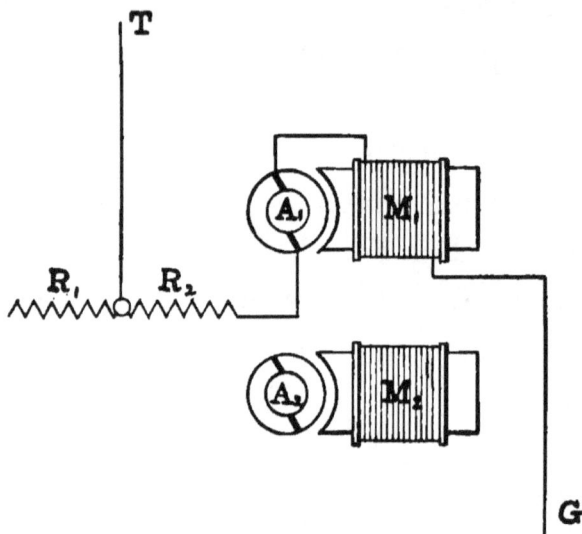

FIG. 77.—CONNECTIONS CORRESPONDING TO SIXTH
WORKING NOTCH OF CONTROLLER.

the speed will be roughly double that
corresponding to the connections in Fig.
75; for, the single motor armature must
produce, roughly, double the C. E. M. F.

of rotation that it produces when it was aided by the motor M_2.

The connections of the seventh working notch are the same as for the sixth, or remain as shown in Fig. 77. This is merely for the purpose of not making the next change too suddenly, requiring the motorman to take a certain time in turning his handle for two notches, so as to avoid abrupt changes in speed.

The conditions produced when the eighth working notch is reached are indicated diagrammatically in Fig. 78. Here the second motor M_2, which was withdrawn from the circuit in Fig. 77, is replaced *in parallel* with M_1, instead of in series; that is to say, the current through the circuit divides, half passing through M_1, and half through M_2. Each motor,

however, must make up, disregarding drop
in R_2, the full pressure of 500 volts between trolley and track, and the speed of
rotation would remain practically un-

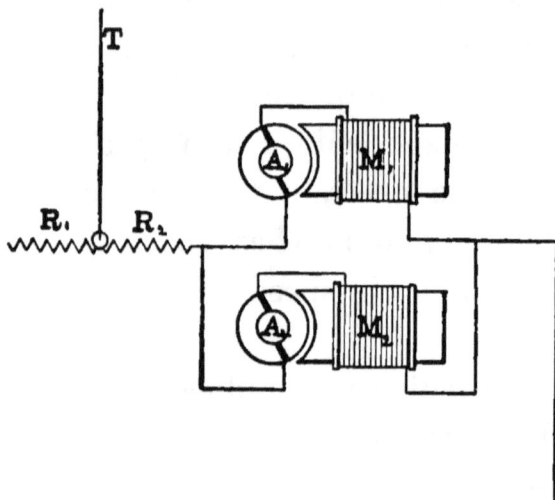

FIG. 78.—CONNECTIONS CORRESPONDING TO EIGHTH
WORKING NOTCH OF CONTROLLER.

changed, except that the current, being
approximately halved through each magnet, the strength of the magnetic field is
weakened, and the armatures have to run

faster to make up the required C. E. M. F., in this weakened magnetism. The speed of the car will, therefore, be greater than in the case represented in Fig. 77.

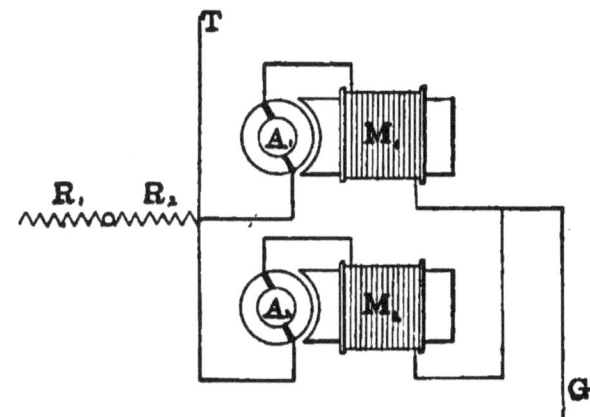

FIG. 79.—CONNECTIONS CORRESPONDING TO NINTH WORKING NOTCH OF CONTROLLER.

The effect of turning the controller handle to the ninth working notch is represented in Fig. 79. Here the resistance R_2, is completely cut out of circuit and the two motors are in parallel as in the last case ; or, as it is sometimes called, are connected *in multiple*. The speed will be increased, owing to the fact that the drop previously existing in R_2, now requires to be made up by the motors alone.

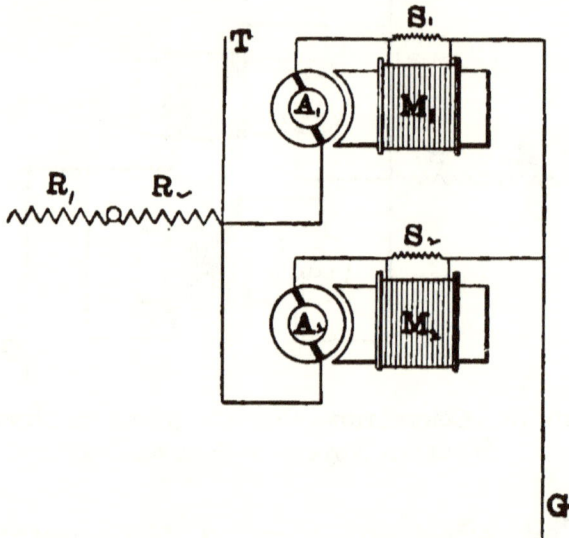

FIG. 80.—CONNECTIONS CORRESPONDING TO TENTH WORKING NOTCH OF CONTROLLER.

Fig. 80, represents the connections corresponding to the tenth and last working

notch ; *i. e.*, the connections for full speed.
Here the only change from the connections of Fig. 79 lies in the restoration of
the shunts around the field magnets, thereby
reducing their excitation and requiring
an increased armature speed in order to
maintain the C. E. M. F.   Each motor, as
before, has to produce, in C. E. M. F. and
drop, the full pressure of 500 volts, and
when the field is weakened, the speed for
a given C. E. M. F. of rotation has to
increase.

It will be observed, therefore, that the
movement of the controller handle through
the successive notches, results in an increasing speed of the car.   Of course
movement in the opposite direction results
in changing the connections in opposite
order of succession ; and, consequently,
slows the car.

There is no definite or precise speed
which corresponds to each notch, since that
will depend upon the load of the car and
the gradient at which it runs. In other
words, it will depend upon the activity
which the motors exert. The lighter the
load for any given notch or set of connections, the faster the motors will run. On
the contrary, an increase of load at any
time, even without touching the controller
handle, will result in a diminution of
speed.

The function of the small handle h_i is to
reverse the direction of current through the
two motor armatures, and, consequently,
to reverse their direction of rotation.
As this cannot be safely accomplished
during the running of the motors, the
handle h, is so arranged mechanically that it
cannot be turned until the controller handle

H, is at the "off position" on the first notch; so that before the car can be reversed the current must first be shut off. This prevents any arcing on the contacts of the reversing cylinder y. All the arcs which tend to form on the contact segments of the large cylinder are extinguished by the action of the magnet M, which is always in the circuit.

At the bottom of the controller are two switches m and n, respectively. These are commonly employed to cut out one of the motors on the car, if by any accident it should become disabled. For example, if the brushes of motor M_1, should fail to make good contact, or give other electrical trouble, that motor can be entirely cut out of circuit. Similarly, by lifting the switch handle n, the motor M_2, can be entirely cut out of circuit. In such a case the car is

operated by the remaining motor, and
only such notches can be used with the
controller handle as will be available for
the operation of that motor.

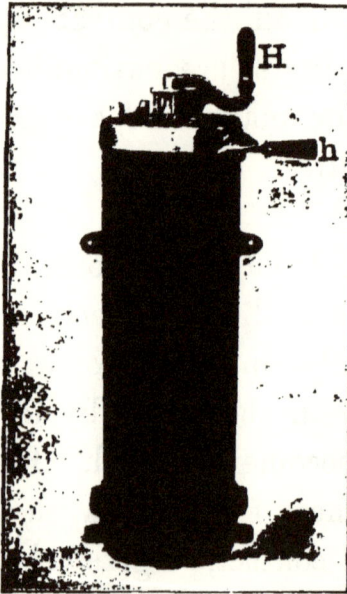

FIG. 81.—STREET CAR CONTROLLER.

Another form of controller is shown
in Figs. 81 and 82. Here, as before, we

have the main controller handle *H*, and a small reversing handle *h*. The method of operation is substantially the same in

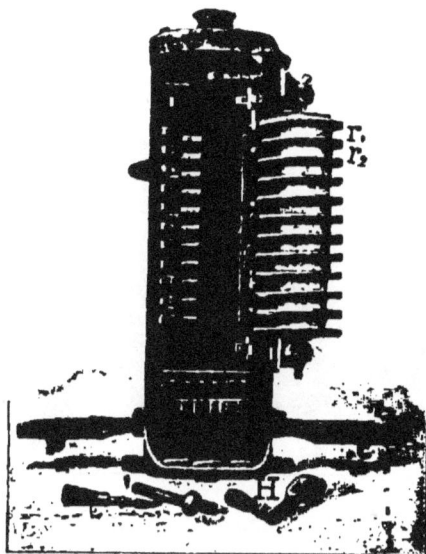

FIG. 82.—CONTROLLER OF FIG. 81 OPENED FOR INSPECTION.

all controllers. In this case, however, no attempt is made to blow out the arcs magnetically when breaking the circuit. Instead of this the arc is caused to occur simultaneously at a number of segments in series, so as to produce a number of small arcs instead of a single large one. This greatly reduces the heat and deflagrating

FIG. 83.—FORM OF CONTROLLER RESISTANCE.

power of the arcs. The contact points at which they occur are renewed from time to time.

Fig. 83, shows a form of resistance employed with the controller represented in Figs. 81 and 82. Here the resistances are formed of strips of sheet iron, wound upon insulating frames, in coils or cylinders, three of which are stowed in the iron box shown, in such a manner as to allow free circulation of air to carry off the heat that may be generated in them. There are four screw terminals t_1, t_2, t_3, t_4, placed on an insulating slab at the top of the case for the wires to connect with.

The controller of a car may be regarded as a complex switch capable of effecting the different connections such as we have indicated. Usually there is one controller at each end of the car. The handle H, is carried from one controller to the other according to the direction in which the car is to be run.

In order to protect the controller or

motors from any excess of current, an *automatic cut-out* or *safety fuse* is employed in the circuit. This consists of a copper wire, of such size that it will melt when the current attains an excessive strength.

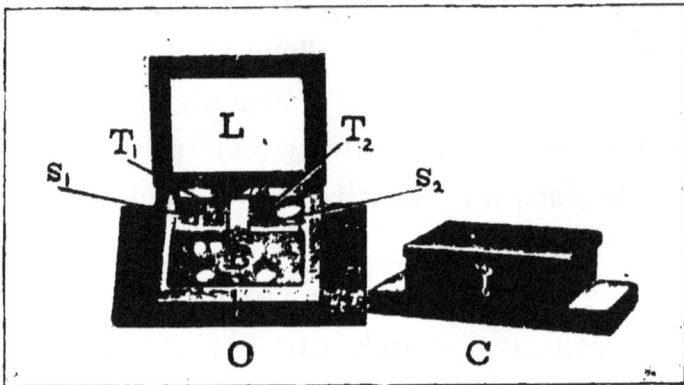

FIG. 84.—FUSE BLOCK.

The wire is enclosed in a box or block called a *fuse block*, placed in a suitable position on the car, usually on the platform overhead, where it can be readily inspected. A form of fuse block is represented in Fig. 84. The block, as it appears when closed, is shown at C, and, as it appears open, at O. A block of hard wood B, carries, secured to its edge, two screw binding posts S_1 and S_2, and tongues T_1, T_2. The clips are permanently in connection with the trolley on one side, and the controller on the other, so that the current has to pass from the trolley through the fuse block by means of these clips. Connection is made between the clips through a wire, usually either No. 12, or No. 14, A. W. G., running around the edge of the block B, and having its extremities clamped under the screws s_1 and s_2. The lid L, of the box, as well as its interior, are lined with asbestos cloth to prevent damage through the melting of the copper fuse.

In addition to the controller and fuse block there is usually added a *canopy*

switch at each end of the car. This switch
is provided for the purpose of permitting
the motorman to turn the current on or off
the car as desired, when, for example, he
wishes to inspect a fuse block or controller, without pulling down the trolley

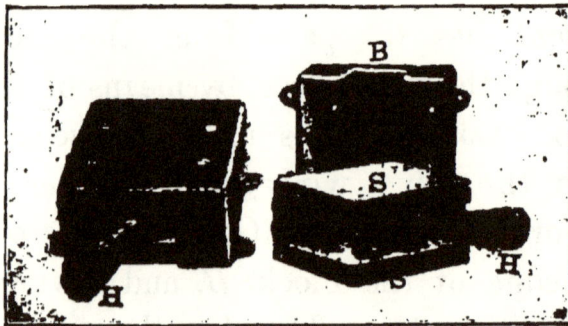

FIG. 85.—CANOPY SWITCH.

pole. It receives its name of canopy
switch from its position beneath the
canopy or roof of the platform.

Fig. 85, shows a form of canopy switch.
A cast iron box *B*, encloses the working

parts and screws up against the canopy. The handle H, projects from this box and can be moved sideways in the slot or groove provided for the purpose. This insulating handle is fastened to a metallic blade which closes a contact with a clip C, thus establishing the main circuit from the trolley to the controller. S, S, are two slotted slabs between which the handle plays.

To protect the motors and apparatus on a street car from electrical discharges produced by atmospheric disturbances; *i. e.*, from lightning discharges, a *lightning arrestor* is usually included in their equipment. A form of lightning arrestor is represented in the accompanying figure 86. Here a cast iron box B, B, with its lid L, L, removed for inspection of its interior, has a pair of marble slabs, the upper

one of which is shown at *M,* clamped
together by screws. A groove runs down
their interior surface, between two metal-

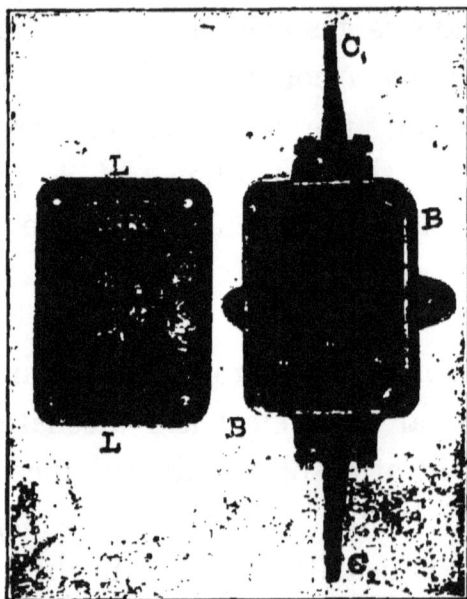

FIG. 86.—LIGHTNING ARRESTOR.

lic pieces c_1 and c_2, in electrical connection
with the leads or insulated conducting
wires C_1, C_2. This groove is black-leaded

in such a manner as to provide a ready path for discharges of very high E. M. F., such as those which accompany lightning discharges, but forms an effectual barrier, or high resistance path, to currents from a pressure of 500 volts. Should a lightning discharge occur between the trolley wire C_1, and the ground or track wire C_2, the dynamo current will be unable to follow this discharge owing to the rapidity with which the heated column parts with its heat to the marble blocks. In other words, the conducting path is chilled so suddenly, after the passage of the momentary high-pressure discharge, that the dynamo current is unable to follow. If this were not effected the high-pressure discharge would establish a very powerful and dangerous arc between trolley and track.

CHAPTER VIII.

THE existing system of trolleys and trolley wires for street railway cars, simple as it seems, has, nevertheless, been the outcome of no little practical development and experience. At the present time the system in almost universal use is the *single-trolley system.* In this system, a current is taken from an overhead wire suspended over the street. After passing through the motors the current returns to the power station, through the track and ground return.

The well known mechanism provided

204

for transferring the current from the trolley wire to the cars, called the *trolley mechanism*, is shown in Fig. 87. As will be seen, it consists of a light steel pole p, called the *trolley pole*, mounted on a base b, called the *trolley base*, and provided at its extremity with a light wheel t, called the *trolley wheel*. The rope r, called the *trolley rope*, is provided for pulling the trolley away from the trolley wire $w\ w$, and for aiding in replacing it.

Simple as the trolley mechanism appears, nevertheless, certain conditions must be satisfied, in order to ensure efficient operation. One of the most important of these is that sufficiently firm pressure or contact be steadily maintained between the trolley and the wire under which it runs. Moreover, this contact must be flexible. The requisite flexibility is obtained both

FIG. 87.—PASSENGER CAR WITH TROLLEY.

by the flexibility of the trolley wire itself, and the mounting or support of the trolley on its base. Means too, are provided for reversing the direction of the trolley pole, so that the car may be driven in either direction. For obvious mechanical reasons the trolley pole always slants away from the direction in which the car moves.

The *trolley wheel*, or *trolley*, is the name given to the revolving part which is supported at the top of the trolley pole, and maintained in rolling friction upon the under side of the trolley wire. Its function is to maintain electric contact with the wire, so as to take from it the current required for the operation of the car. One form of trolley wheel is seen in Fig. 88. As here shown, it consists of a light wheel *W*, usually of gun metal, supported in a *frame* or *harp H*, and running

freely upon a spindle, not shown in the figure, passing through both harp and wheel.
The grooved form given to the wheel not
only serves the purpose of ensuring a

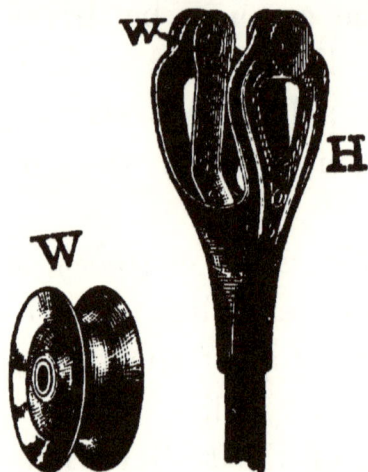

FIG. 88.—TROLLEY WHEEL AND HARP.

more extended rolling contact surface
with the wire, but also serves to prevent
the trolley from slipping off the wire. The
spring *w*, pressing against the face of the
trolley, maintains good electric contact
between the wheel and an insulated wire

which passes down through the trolley
pole to the car.

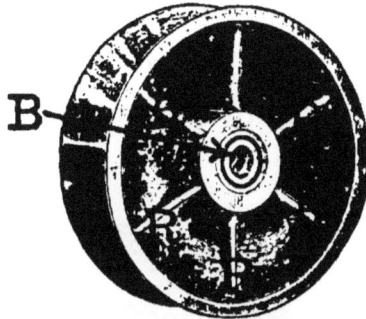

FIG. 89.—FORM OF TROLLEY WHEEL.

Various forms of trolley wheels have
been devised. It is essential that they

FIG. 90.—FORM OF TROLLEY WHEEL.

shall be as light, rigid and freely running
as possible. For this purpose, special
attention is paid to their lubrication,
which is usually effected by employing a
bushing of graphite, or other lubricating
material.

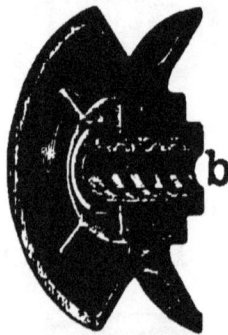

FIG. 91.—SECTION OF TROLLEY WHEEL, SHOWING
LUBRICATING BUSHING.

As an illustration of some of the various
forms of trolley wheels those shown in Figs.
89 and 90 may be taken. It will be observed that these wheels are ribbed, so as
to ensure strength combined with light-

ness. Moreover, should the rim of the wheel wear out and drop off during a trip, the trolley wire will still be gripped by the ribs R, R. The bushing of lubricating material is seen at b, Fig. 91, which shows a section through the wheel of

FIG. 92.—LUBRICATING BUSHING.

Fig. 89. Here the lubricating bushing B, is seen in place at the centre of the hollow wheel. Fig. 92 shows a form of bushing ready for insertion.

At times during winter, when the trolley wire is covered with sleet, some difficulty is experienced in taking off the

current, ice being practically an insulator.
Various devices have been suggested to
avoid this difficulty. A form of trolley
wheel, which assists in clearing sleet from
the wire, and allows the fragments of ice

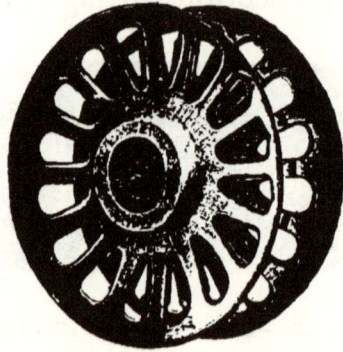

FIG. 93.—SLEET-CUTTING TROLLEY WHEEL.

to escape through the sides of the wheel
is shown in Fig. 93.

The trolley pole is in almost all cases a
steel tube, tapering toward the top. Its
lower end is mounted on the trolley
frame or base. Springs are connected

between the base and the pole in such a
manner as to maintain the pole in contact
with the wire, with a nearly uniform pressure, under all conditions of dip or devi-
ation of the trolley wire. Various trolley

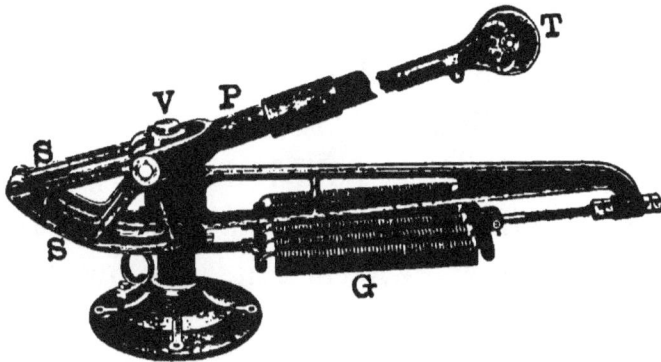

FIG. 94.—TROLLEY BASE.

poles and bases have been employed. A
well known form of trolley base is
shown in Fig. 94. Here the pole P,
terminates in a fork attached to a pair
of sectors S, S, forming a frame, capable
of revolving about a vertical axis V,

so as to accommodate the pole and trolley wheel to turns or curves in the track and trolley wire. The six spiral springs G, maintain a tension upon these sectors tending to force the pole P, upwards. This tension can be altered by the screw adjustment behind the springs. In order to be able to use the trolley when the direction of the car is reversed, the pole is first pulled down from the trolley wire and then swung around the vertical pivot V, when it is allowed to re-engage with the wire in the opposite direction.

Another frame and pole called the *Boston trolley* apparatus is represented in Fig. 95. The wooden frame $FFFF$, is screwed to the roof of the car. It carries a spindle, working on a horizontal axis and bearing the pole P, at its centre.

Eight spiral springs G, G, maintain the requisite tension upon the pole under the screw adjustment s s. Two smaller spiral springs g, g, are provided for supporting the pole in the vertical plane, and

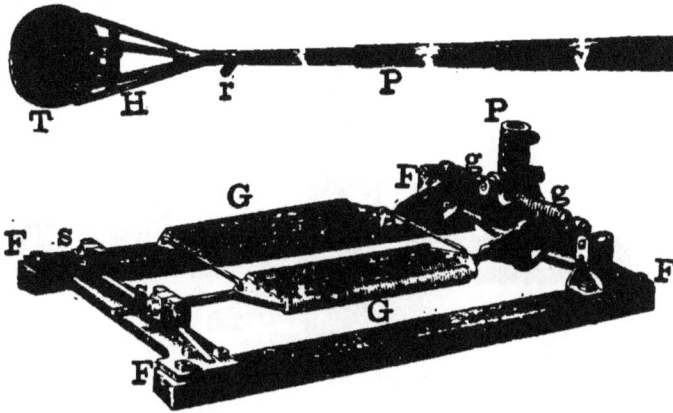

FIG. 95.—BOSTON TROLLEY BASE AND POLE.

help to keep it from leaving the wire. T, is the trolley; H, the harp; r, the attachment; and P, the pole.

A simple form of trolley base is shown in Fig. 96. Here the pole P, is supported

in a fork *F*, carrying two lugs *l, l*, connected on each side of the pole by rods to
the extremities of stout spiral springs.
The effect of these springs is to maintain
the trolley pole vertical under ordinary
circumstances, and, when the pole is

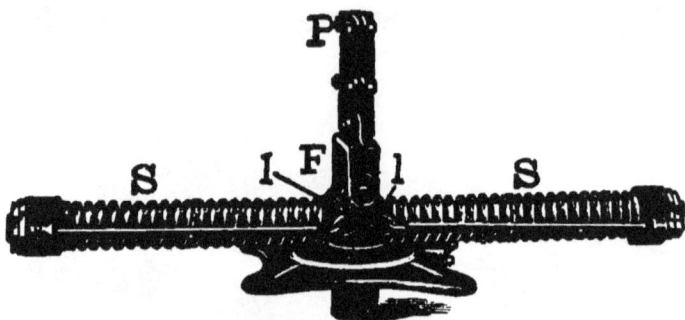

FIG. 96.—FORM OF TROLLEY BASE.

pulled down, it tends to return to the
vertical position by the compression of
one spring and the distension of the other.
The pole and springs together can swing
around the vertical axis upon which they
are mounted so as to accommodate the
trolley to curves.

FIG. 97.—TROLLEY POLE AND BASE.

Other forms of trolley poles and bases
are shown in Figs. 97 and 98. The mech-

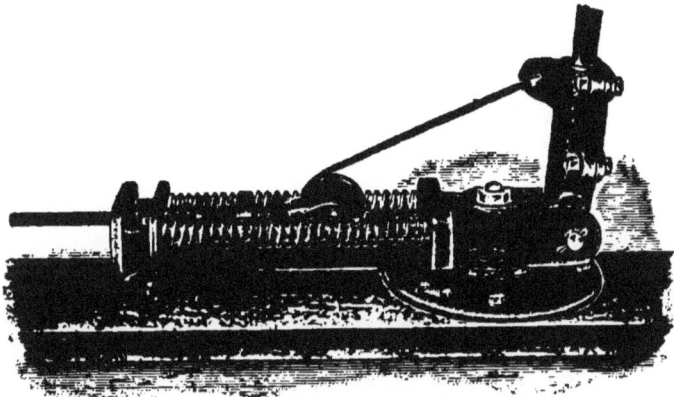

FIG. 98.—TROLLEY BASE.

anism is sufficiently clear in each case
to be understood by a mere inspection.

The angle which the trolley pole makes
with the roof of the car, under ordinary
circumstances, is about 40°. The trolley
wheel is ordinarily pressed upward with
a force of about 30 pounds weight against
the wire.

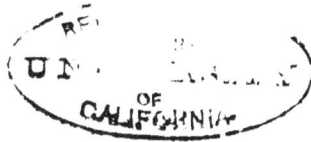

CHAPTER IX.

TROLLEY LINE CONSTRUCTION.

THE poles which support the trolley wire over the track are either of wood or of iron. In the country, wooden poles are frequently employed, while in cities iron poles are preferred. The methods most frequently used for supporting the trolley wire are either by the use of *span wires* or by *brackets*. Span-wire construction requires poles in pairs, on opposite sides of the street, while bracket suspension only necessitates a single line of poles even for double tracks. Where, however, bracket poles are used for double tracks they are open to the objection of requiring to be

placed in the middle of the street, thus
tending to obstruct traffic.

Fig. 99, shows the *span-wire system*,
with two iron poles, *P*, *P*, made of three

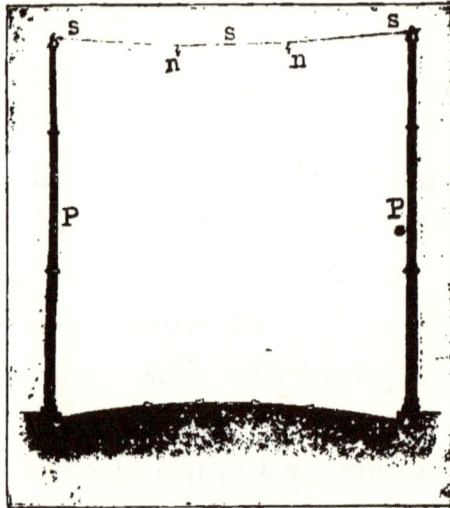

FIG. 99.—SPAN-WIRE SUPPORT.

tapering lengths of iron tube. *s, s,* is the
span wire, commonly of No. 1 A. W. G.
iron wire; *n, n,* are the insulators sup-

ported on the span wire, and in their turn supporting the two trolley wires over their respective tracks. The poles are

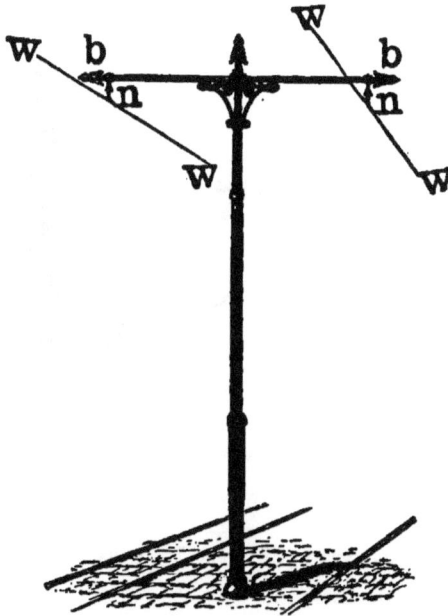

FIG. 100.—BRACKET POLE FOR DOUBLE TRACK.

commonly 27 to 30 feet long, and are buried to a depth of 6 feet, being usually set in concrete. For span-wire construc-

tion, the poles are commonly set slanting
from the tracks so as to enable them
better to stand the strain of supporting
the trolley wires.

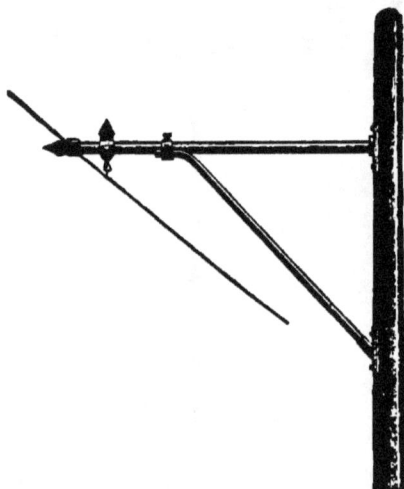

FIG. 101.—SINGLE-TRACK BRACKET SUPPORT.

The poles for the *bracket-support system*
are always set vertically and midway between the tracks. Such a pole is shown
in Fig. 100. Here *b, b,* is the bracket arm
and *n, n,* the insulators suspended there-

from, supporting in their turn the trolley
wires *w, w*, and *w, w*.

Forms of *single-track bracket suspension*
are shown in Figs. 101 and 102. The

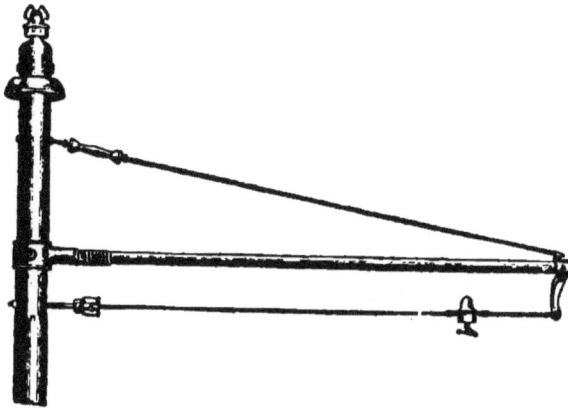

FIG. 102.—SINGLE-TRACK BRACKET SUPPORT.

poles are set about 120 feet apart; *i. e.*,
about 45 per mile with bracket suspensions, or 90 per mile for span-wire suspension.

In order to attach the span wires to the

iron pole, *iron clamps* are employed, generally of the form shown in Fig. 103. When the clamps are in position facing each other on opposite sides of the street, the span wires are stretched between them

FIG. 103.—POLE CLAMP.

under considerable tension, depending upon the weight of trolley wire, but 500 pounds weight is a fair tension. Where only a single span wire crosses the street, it is often stretched between insulators at the top of the poles, as shown in Fig. 99.

Since, of course, the trolley conductor is

an uninsulated wire, *guard wires* are often employed to prevent damage from contact with bare telegraph or telephone wires, which would thereby become connected with a pressure of 500 volts. Guard wires are of two kinds; viz., *span guard wires*, which cross the street immediately above the span supports over the trolley wire, and *running guard wires*, which run parallel with and immediately over the trolley wires to receive and intercept any wire falling from above. The relative position of guard and suspension wires is illustrated in Fig. 104. *P*, *P*, are opposite poles, *c*, *c*, the pole clamps, *s*, *s*, the suspension span wire, and *g*, *g*, the guard span wires. The trolley wires are always suspended from the lower wire *s*, *s*, and guard wires are usually suspended over the trolley from the upper span *g*, *g*.

We will now turn our attention to the
devices adopted both for supporting the
trolley wire from the suspension span wire,
and for enabling the trolley to be stretched

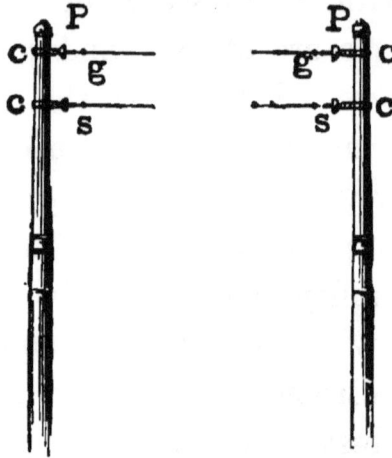

FIG. 104.—POLES WITH GUARD AND SUSPENSION SPAN
WIRES.

tightly. It is necessary not only to support the wire rigidly and to insulate it from
the span wire, but also to employ devices
for this purpose that shall be as small and
sightly as possible. The simplest way to

support a trolley wire from a span wire is by means of a *trolley ear* or *insulator*. Such a form of ear or insulator is shown in Fig. 105. *e, e,* is a metal casting, called

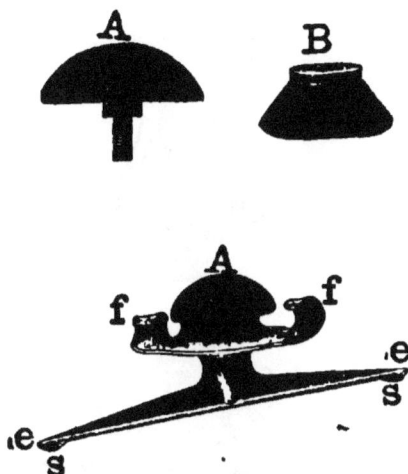

FIG. 105.—STRAIGHT-LINE SUSPENSION, AND TROLLEY EAR AND INSULATOR.

the ear. It is furnished with a narrow edge *s, s,* having tips which are bent and soldered over the trolley wire, which lies in a groove extending under the entire length of the ear. *f, f,* is the body of the suspen-

sion, having two flanges at its extremities
as shown. The suspension span wire lies
in these flanges and around the head of the
insulator. The insulator is made in two

FIG. 106.—TROLLEY EAR AND SUSPENSION.

parts, *A* and *B*, shown separately above,
A, being an insulating cap and *B*, an insulating cone. These two parts are screwed
together and grip the body between them.
Fig. 106 shows another form of street line

suspension and ear, differing from the former merely in details of construction. The outer iron cap C, has the cover v, screwed down upon it in such a manner as to enclose the insulating tube t. This insulating tube encloses in its turn the bolt which is screwed into the ear. The sus-

FIG. 107.—DOUBLE-CURVE SUSPENSION.

pension span wire is gripped tightly between the flanged projections f, f, of the body and the outside of the iron cap c.

A great variety of *line suspensions* are employed. Fig. 107 shows a common form called a *double-curve suspension*, named

from the two lugs of the hood or cover.
On the insertion of this form of suspension,
the span wire has to be cut and the two
ends fastened into the rings *r*, *r*. In other
respects the suspension is practically the
same as that shown in Fig. 106. The

FIG. 108.—SINGLE-CURVE SUSPENSION.

double-curve suspension possesses the advantage that all the tensions exerted upon
it, with the exception of that produced by
gravitation, are exerted in the horizontal
plane; that is to say, the span wire pulls
sideways upon it in almost the same plane as
the tension of the trolley wire lengthways.

Another form of suspension, called the *single-curve suspension* is shown in Fig. 108. This suspension is introduced at curves in the track or line where only a

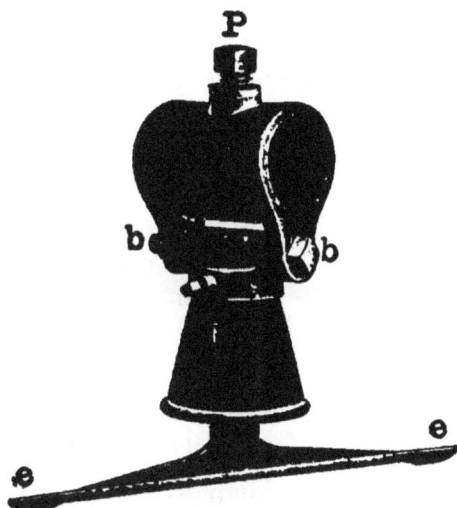

FIG. 109.—BRACKET SUSPENSION EAR.

single pull is exerted on the trolley wire, instead of requiring a span.

A form of *bracket suspension ear*, is shown in Fig. 109. Here the cylinder *C*, is clamped firmly by the screw clamp P, upon a bracket arm, while from the cylinder is supported the insulator I, and ear e e, upon the bolt b b.

When two lengths of trolley wire have to be connected together, the connection is always made at an ear, or point of support.

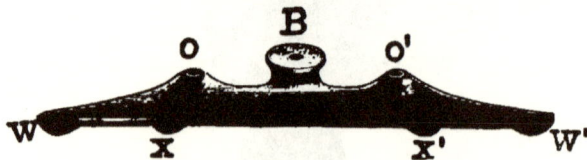

FIG. 110.—SPLICING EAR.

Such an ear is for this reason called a *splicing ear*. A form of splicing ear is shown in Fig. 110. The two ends to be connected are brought respectively to the ear at w and w', under the grooves to x and x', and then through the holes in the ear at the openings o and o'. The wires are then soldered in at w, x and o. The ear is bolted to its supporting insulator at *B*.

Instead of soldering the ends of the wire in a splicing ear they may be clamped in a device called an *automatic ear*, shown in Fig 111.   Here the two wires are laid

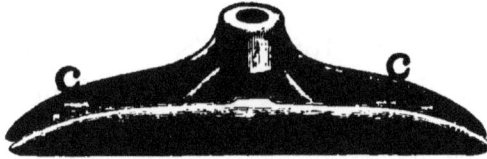

FIG. 111.—AUTOMATIC OR CLAMP-SPLICING EAR.

in the jaws of the clamp at *C, C.* The jaws are then pressed together and secured by a bolt.

The necessity for maintaining a taut trolley line, so as to ensure a good and continuous contact with the trolley wheel, requires that the line be *anchored* about every 1,000 feet. An *anchor-strain ear* is shown in Fig. 112. *Strain wires* are

FIG. 112.—ANCHOR-STRAIN EAR.

attached to the lugs *a* and *b*, and are made fast, through insulators, to equidistant poles as shown in Fig. 113. The insula-

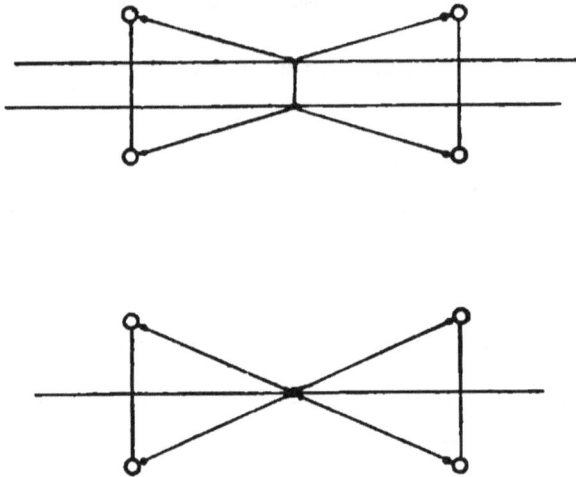

FIG. 113.—ANCHORING FOR SINGLE AND DOUBLE TRACK.

tors which are employed for this purpose are called *strain insulators*, and are of various forms.   A common form is shown in Fig. 114.   The two lugs are cast into a spherical insulating mass.

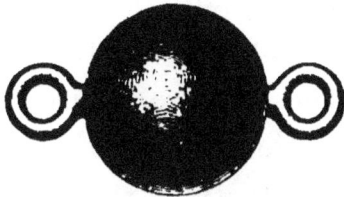

FIG. 114.—STRAIN INSULATOR.

Trolley wire insulators have two functions to fill; namely, a mechanical function; *i. e.*, in providing an adequate support, and an electrical function; *i. e.*, as an electrical insulator.   In order to be sufficiently strong, suitable material must be employed and so arranged as safely to support the stresses exerted upon it.   From an electrical point of view, the insulation afforded by an insulator is never that of the material of which the insulator is formed, and is always, in practice, the insulation of the surface. That is to say, the electric leakage, which takes place through an insulator, is practically all over the surface of the insulator, scarcely any passing through the substance of which it is formed. The condition of the surface, therefore, greatly affects the efficient action of the insulators; for, if dirty or dusty, a thin film of moisture will entail a considerable electric leakage. Assuming the same surface conditions, a spherical insulator, such as that shown in Fig 114, would permit considerably greater leakage than a cup insulator of the type shown in Fig. 106, especially in wet weather. The electric leakage, however, which can be permitted on a trolley system is far in excess of that which can be allowed on a telegraph or telephone circuit; since, if the total line leakage gave rise to a loss of activity amounting to 1 KW, which would represent a total leakage current of 2 amperes under a pressure of 500 volts, or a total insulation resistance of only 250 ohms; the cost of this would be one or two cents per hour. The insulation of trolley systems usually averages from 2,000 to 100,000 ohms to the mile according to the weather.

When a trolley road branches, it is necessary to branch the trolley wire. This is accomplished with the aid of a device, called a *trolley frog*. Fig. 115, shows three forms of trolley frogs. At *A*, is a *V-frog*, or simple *two-way frog*, in an inverted position, so as to show the guides. *a*, is a metallic guide on the side of the single track, and *b* and *c*, are the two guides on the side where the road bifurcates. When the car has to be driven, say from *a* to *b*, the rails on the track are so switched as to carry the car in that direction, and the trolley follows from the

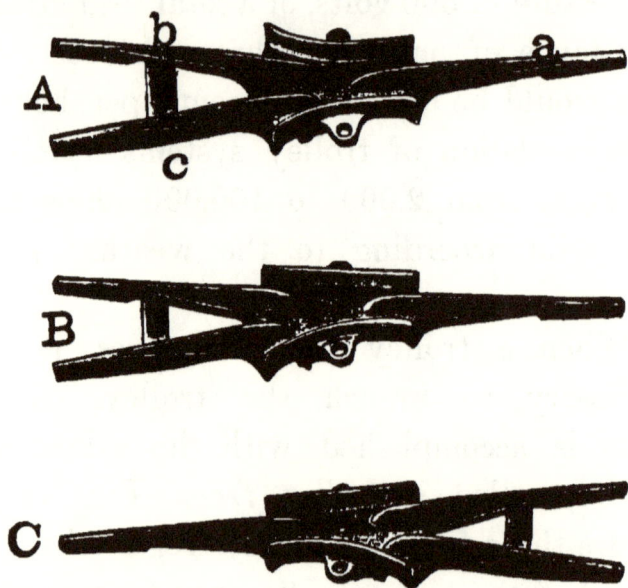

FIG. 115.—TROLLEY FROGS.

guide *a*, to the guide *b*.   During the passage from the guide *a*, to guide *b*, the trolley wheel will either maintain contact with the line through its metal frame, or may make a momentary flash at the point of crossing. *B*, shows an inverted *right-hand frog* and *C* an inverted *left-hand frog*. Where a line divides into three branches special frogs, called *three-way frogs*, have to be employed.

FIG. 116.—TROLLEY CROSSING.

At the intersection of two streets where trolley wires necessarily cross each other, the crossing is effected through the medium of a device similar to a frog, and called a *trolley crossing*. Forms of trolley crossings are shown in Fig. 116. *A*, is a *right-angle crossing*, and *B*, an *acute-angle crossing*. The trolley wires are soldered in the groove over the four guides, and as a result, the trolley wheel has to drop slightly at a crossing to pass beneath the guides. Special forms of crossings are employed when it is desired to insulate the two crossing trolley wires from each other.

Trolley wires are made in all sizes from No. 4 A. W. G., with a diameter of 0.204″, to No. 000 A. W. G., with a diameter of 0.410″. The commonest size is No. 0, of 0.3249″ diameter. The material is usually hard-drawn copper, although alloys are occasionally used. A No. 0, hard-drawn copper wire will safely bear a tension of 2,500 lbs. weight, and usually breaks at a tension of 5,000 lbs.

weight. A hard-drawn copper wire of this size has a resistance of, approximately, 0.52 ohm at 60° F., its resistance being about 2 1/2 per cent. in excess of the resistance of the same size wire in soft copper, whereas silicon-bronze wire has sometimes about 2 1/2 times the resistance of the same size of soft copper wire.

CHAPTER X.

TRACK CONSTRUCTION.

It is frequently a matter of surprise that the installation of a trolley road is almost invariably attended by the reconstruction of the track. The necessity for this reconstruction is to be found in the fact that electric cars are much heavier than ordinary horse cars, and contain running machinery which is liable to injury from excessive jolting. This liability to injury from a weak and inferior track is increased by the greater speed at which electric cars run. Moreover, in a badly constructed track difficulty is experienced in maintaining an efficient running contact

242

between the trolley and the trolley wire. For these reasons the construction of the roadbed and track requires careful attention.

In cities more care and expense are naturally taken with both line and track

FIG. 117.—TRACK CONSTRUCTION.

construction than in the open country, but the tendency is towards the employment of a steel girder rail weighing 90 lbs. per yard. These rails are laid directly on wooden sleepers to which they are spiked. This construction is shown in Fig. 117, where the girder rails R, R, are spiked to

the sleeper *S, S,* and are also bound together by the tie rod *T, T,* the roadbed
being paved in this case with Belgian
blocks. The rails are laid with their ends
close together, no difficulty having been
experienced from expansion in summer

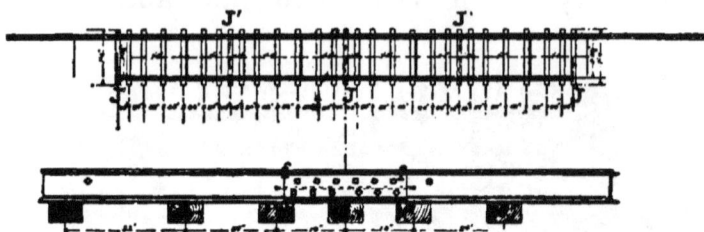

FIG. 118.—TRACK AND SLEEPERS, SHOWING METHOD OF
BREAKING JOINTS.

time. It is common to break these joints
so that the joints of the rails on one side
of the track shall come opposite to the
middle of the rail on the opposite side.
This is represented in Fig. 118, where
J, J, J, and *J', J',* show the relative
positions of the joints of each rail. The

sleepers in this case are also so distributed as to be closer together near the joints, as shown. *f, f,* is the fish-plate with twelve bolts which pass through the rail and are screwed up against a similar fish-plate on the other side of the rail.

With the use of a ground return it is necessary to ensure as intimate a contact between the rails as possible, so as to secure a continuous metallic path and to lessen the resistance that would otherwise be introduced into the circuit. Mere contact of the ends of the rails with their connecting fish-plates is not sufficient, since rust at this surface produces a very considerable resistance. In order to avoid this, various methods of *bonding* the rails have been proposed. This is attempted in a variety of ways, but the object is always to secure a permanent metallic connection between successive rails. One of these *rail bonds* is represented in Fig. 119. To use this bond the rails are drilled close to the fish-plate and a bent copper rod of the shape shown at *A*, has its two ends pressed into the holes, one end in each rail. A section

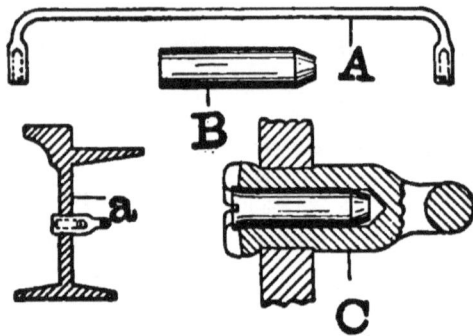

FIG. 119.—CHICAGO RAIL BOND.

of the rail with the end of the copper rod projecting through it is shown at *a*. The plug *B*, is then driven with the hammer into the opening of the rod so as to wedge it tightly into the iron rail. A cross-section of the rail, rod and plug is shown at *C*.

A somewhat similar method of effecting a rail bond consists in the use of stout copper wire in place of the copper rod. Here the wire is passed twice through holes in the rail each side of the fish plates and copper wedges are driven in so as

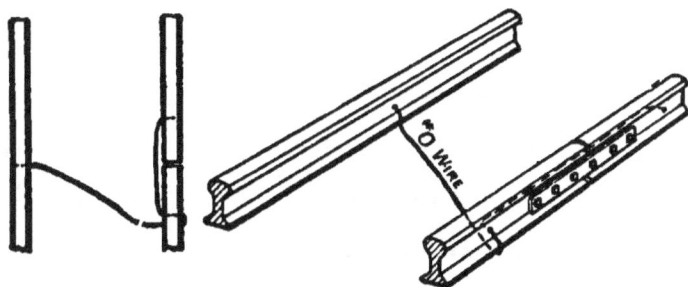

FIG. 120.—WIRE RAIL BOND.

completely to wedge the wire against the metal rail. At intervals this wire is led directly across the track and enters into a bond with the other rail, thus effectively connecting the two rails together. A wire bond of this character is shown in Fig. 120.

The most efficient bond from a purely electric point of view is the *welded rail bond* obtained by welding the rails together. For this purpose a very powerful electric current is passed through the ends of the rails, and pieces of iron, called chucks, which are used in place of the fish-plates. When completed, this joint is as solid and strong as the rest of the rails, thus affording a practically continuous iron rail, and therefore a continuous return circuit. Another method of accomplishing the same result consists in pouring melted cast iron around the ends of the rails after cleaning them, and so effecting a solid joint. Although success has not yet been perfectly obtained with continuous rails, yet it would appear that the stresses produced by expansion and contraction in a uniform continuous rail are well within the limits of the elasticity of the steel.

CHAPTER XI.

ELECTROLYSIS.

WHEN an electric current is sent through a vessel containing ordinary tap water, the passage of the current is attended with the decomposition of the water into its constituent elements, oxygen and hydrogen. These elements are liberated, in the gaseous state, only at the points of entrance and exit of the current from the water, the hydrogen being liberated where the current leaves the water, and the oxygen where the current enters the water. If the conducting surface at which the current enters is oxidizable like iron, copper, lead, zinc, and nearly all ordinary metals

249

it becomes corroded or oxidized, while a similar metal surface or electrode provided for the exit of the current from the water is unaffected, the hydrogen being usually disengaged in bubbles. Decomposition effected in this manner, by an electric current, is called *electrolytic decomposition*, and the corrosion of metals in liquids in this manner is called *electrolytic corrosion*.

The earth or ground is only capable of acting as a return circuit by virtue of the moisture which is practically always present. Consequently, in all cases where the ground-return circuit is used, the metallic surfaces by which the current enters and leaves the ground are liable to electrolytic action. Where the current leaves the metallic conductors to enter the ground, or the moisture within the ground, there will be electrolytic corrosion, but where the current enters a metallic conductor on leaving the ground there will be no electrolytic corrosion, although there may be a liberation of hydrogen. On the contrary, there will be an electric protection afforded the metal, at such points—the

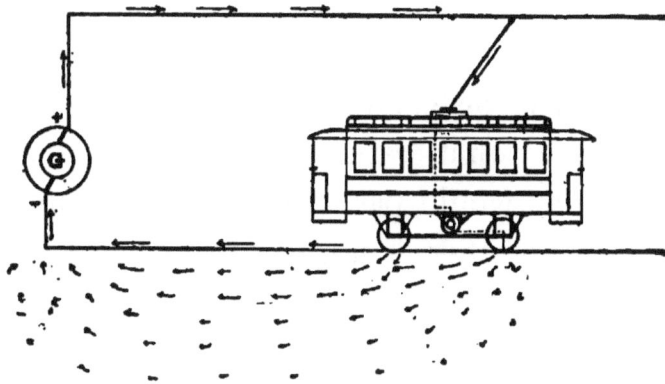

FIG. 121.—SIMPLE TROLLEY CIRCUIT.

oxidation being less than that of similar metal, exposed to ordinary conditions in the absence of electric currents.

The simplest condition of a trolley system is represented in Fig. 121. Here the

generator G, has its positive pole connected to the trolley, that is, the current
enters the trolley from the generator, passes
through the car motors, and returns to the
generator, partly by the track and partly
by the ground; *i. e.*, the water in the
ground, as a supplementary or auxiliary
conductor. If the track had no electric
resistance, or conducted perfectly, all the
current would return through the track
and none would pass through the ground.
If, on the other hand, the track were disconnected at some point, for instance at
each rail joint, then its resistance would be
indefinitely great and practically all the
current would pass through the ground.

The better the electric conditions of the
rail bonds, and the lower the resistance of
the track, the greater will be the proportion of the current which will pass

through the track and the less the proportion which will pass through the diffused
circuits in the ground. Where the current
leaves the rails on the track, to enter the
ground, there will be corrosion or oxidation
of those rails, but where the current returns from the ground to the track, or other
buried metal at the power house connected
with the generator, there will be no corrosion, and even a tendency to prevent corrosion.

When electrolytic corrosion takes place
the amount is perfectly definite. One
coulomb of electricity passing through
water will dissolve 0.000,002361 lb. of
lead electrode, and 0.000,000,6393 lb. of
iron electrode. Since an ampere is a rate
of flow of one coulomb-per-second, a current strength of one ampere will dissolve
0.000,002361 lb. of lead per second, or

0.000,000,6393 lb. of iron per second, and therefore, if an ampere be steadily maintained for one year it will dissolve by corrosion 74.46 lbs. of lead and 20.16 lbs. of iron. If the current be increased to ten amperes, the amount of lead or iron corroded will be ten times as great, the chemical action being directly proportional to the quantity of electricity which is passed.

In the case of Fig. 121, corrosion will occur over the surface of the track where it lies in contact with moist earth. The corrosion will not be uniform, but will proceed faster at some points than others, the rate of corrosion depending upon the distribution of current over its surface ; i. e., on the local facility with which the current escapes into the earth. The total amount of electrolytic corrosion will depend only on the total quantity of electricity, in ampere-hours or coulombs passing from the metal.

If, however, the generator has its negative pole connected to the trolley wire, and its positive pole connected to the track, the electrolytic conditions will be reversed; for, the current will now leave the metallic surfaces for the moist ground in the vicinity of the power house, and there the corrosion will take place to an aggregate amount depending entirely upon the total quantity of electricity passing into the ground. There will now be no corrosion where the current re-enters the track.

Were the corrosion which occurs with street car systems limited to the track, the consequences would not be so serious, but in cities the corrosion affects the metallic

masses of the gas and water pipes, and
their corrosion may lead to serious damage.
Fig. 122 diagrammatically represents a
street car system in which the positive pole
of the generator is connected to the trolley,
and the negative pole to the track. This

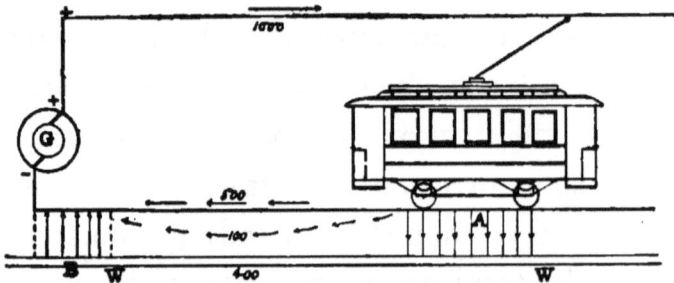

FIG. 122.—DIAGRAM OF TROLLEY SYSTEM IN NEIGHBOR-
HOOD OF BURIED PIPE. NEGATIVE POLE GROUNDED.

case differs from that of Fig. 121, only in
the fact that a system of water pipes, *W*,
W, is supposed to lie in the vicinity of the
track. If we suppose that a current of
1,000 amperes is steadily flowing from the
generator through the car motors, 500

amperes or half the current may return directly to the generator through the bonded track, 100 amperes may return through the ground, escaping from the track at more distant points and returning to it in the neighborhood of the station, while the balance, or 400 amperes, may find its way into the good conducting path presented by the system of water pipes, entering it in the distant areas and leaving it in the vicinity of the power house.

Under the circumstances above mentioned, there will be electrolytic action at A, where the current leaves the track, and at B, where it leaves the water pipe. The area of B, will be a comparatively narrow one, and, consequently, the rapidity of corrosion will be comparatively great, since 400 amperes maintained day and night, represents a total corrosion of roughly

8,000 pounds per annum spread over a comparatively small area. If we connect the water pipe system with the generator's grounded terminal, as shown by the dotted lines, we reduce the quantity of electricity which leaves the surface of B, through the ground, since it will largely pass directly through the new connection. By this means the electrolytic corrosion of the water pipes will be diminished.

If the negative pole of the generator be connected to the trolley and the positive pole be connected with the track, as shown in Fig. 123, then, all other things remaining the same, there will be corrosion at A and B; namely, at the portions of the water pipe remote from the power house and at the portions of the track near it. In this case, however, the area of water pipe over which the corrosion takes place is

more extended, and, consequently, the
amount of corrosion on any one length of
pipe in the district will be correspondingly
less.

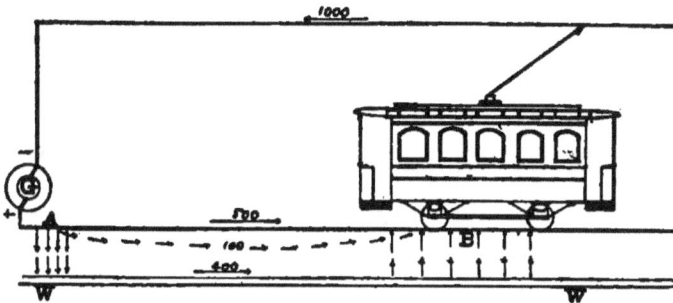

FIG. 123.—DIAGRAM OF TROLLEY SYSTEM IN NEIGHBOR-
HOOD OF BURIED PIPE. POSITIVE POLE GROUNDED.

There are, therefore, two methods of
dealing with the dangerous influences of
electrolytic corrosion upon neighboring
metallic pipes. The first is to ground the
positive pole of the generator or generators
at the power house, and so spread the corrosion over a large area of pipe distant

from the power house, trusting to the enlarged area and the slowness of corrosion to avoid serious effects. In this case there is no advantage to be gained, so far as avoiding corrosion is concerned, by directly connecting the water pipe system with the grounded generator terminal. In fact there will be an advantage in avoiding such connections. The second method is to ground the negative pole of the generator at the power house, as in Fig. 122, so as to bring the area of corrosive action within the neighborhood of the power house. If this course be adopted it becomes important to protect this area by not only connecting the pipes with the grounded generator terminal, but also by securing good electric connections between the track and the grounded terminal of the generator through bonding and ground feeders.

Whichever method be adopted the use of ground ·feeders, rail welding, and efficient bonding necessarily reduces the danger of corrosion by offering a better

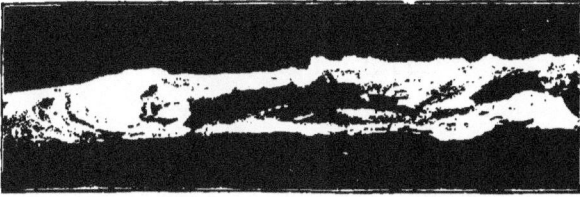

FIG. 124.—IRON PIPE CORRODED BY ELECTROLYSIS.

metallic conducting path to the return current. Fig. 124 represents a piece of pipe destroyed by the influences of electrolytic corrosion.

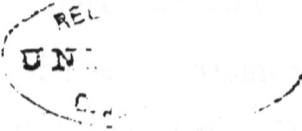

CHAPTER XII.

SWITCHBOARDS.

IF we trace the trolley wires of any street car railway system we will find them to form an interconnected network maintained at, approximately, 500 volts pressure relatively to the track. From this network the feeders pass to the power house, either suspended overhead on poles and insulators, or underground through lead covered cables placed in suitable conduits. Tracing these feeders to their origin we will find them terminating at what is called the *switchboard*. The use of the switchboard is to enable the attendant at the power house to learn at a glance the electric condition of the system, and also to enable him to control or modify the electric condition with swiftness and convenience. To this end the switchboard is provided with a number of electric measuring instruments, called respectively *voltmeters*, for measuring the electric pressure in volts, and *ammeters*, for measuring the electric current in the various circuits in amperes.

Fig. 125, shows a form of *railroad switchboard* intended for use with three separate dynamo generators and three separate feeders. This switchboard consists of seven vertical panels formed of marble, a good insulator. The three panels on the right hand are *feeder panels*, and a generator is connected to and controlled by each. The central panel is a *total-current* and *pressure panel*, for measuring the entire current supplied to the three

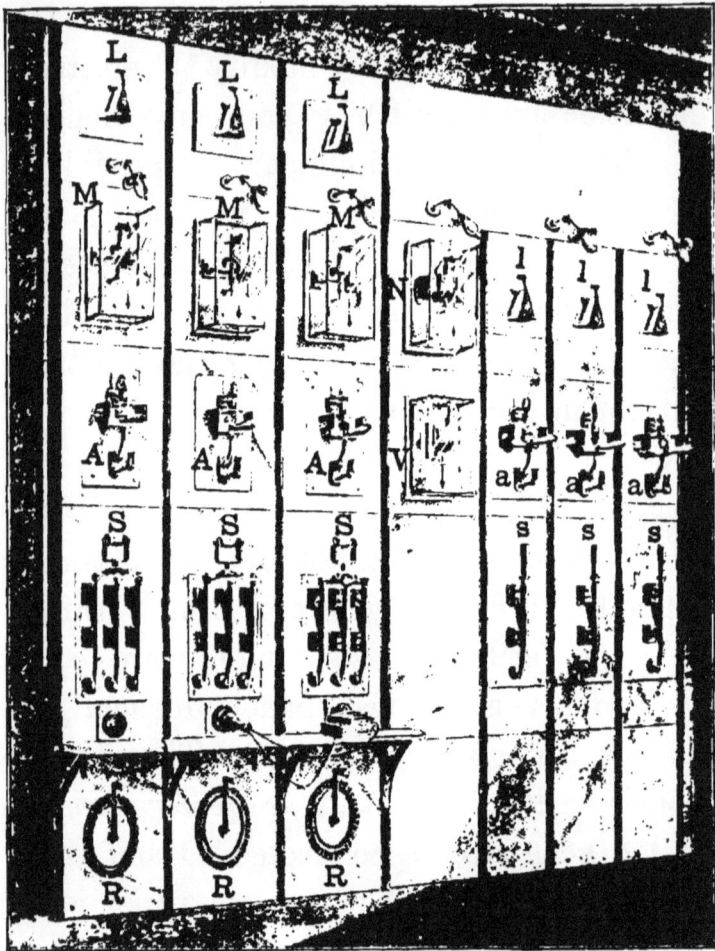

Fig. 125.—Switchboard for Railway Power House.

feeder panels, and the main pressure of the power house. *S, S, S,* are the three *generator switches,* consisting each of three metallic knife blades maintaining connection between metallic clips. In the position shown, all three switches are closed and all three generators are at work together. Beneath the generator switches are *rheostat boxes, R, R, R,* for controlling the current supplied by each respective generator. *A, A, A,* are *automatic circuit-breakers,* which are so arranged that the current, supplied by their respective generators, passes through stout coils or spirals of copper rod, so that when this current strength becomes dangerously great, indicating an overload upon the generator, the magnetic action of the spirals releases a lever, which under the action of the spring flies back and breaks the circuit. *M, M, M,* are three ammeters, each in circuit with its respective generator, so that the pointer or index shows at a glance the current strength and, therefore, the load upon that generator. *L, L, L,* are *lightning arrestors*, intended to carry to ground any discharges due to lightning, thus avoiding damage to the system. Turning to the feeder panels, *s, s, s,* are the three *feeder switches.* On closing one of these switches the particular feeder which supplies it is connected with the generator or generators, which may be in use, so that if all three of the switches shown be opened, the the entire load will be taken off the generators, even though these be maintained running. *a, a, a,* are *automatic feeder circuit-breakers,* similar in their action to those already alluded to at *A, A, A. l, l, l,* are lightning arrestors, connected to each feeder, similar to those at *L, L, L. N,* is the main ammeter, supplied by all three generator ammeters, M, M, M, to-gether, and supplying in its turn, the various feeders. V, is the voltmeter showing the pressure between generator terminals at the station in volts.

The automatic cut-outs A, A, A, and a, a, a, are constructed as shown on a larger scale in Fig. 126. The current sup-plied by the generator passes from the clip P, with its attached carbon plate N, across the metal frame of the switch H, to the opposite metal clip P', and its attached carbon plate N', thence by the terminal A, through the three turns of the metallic coil or spiral C, to the terminal B, from whence it passes to the line. On lifting the handle H, into the position shown on the left hand, a metallic connec-tion is established between the clips, and the switch is kept in position by a detent.

The current passing through the three
turns of the coil *C,* magnetizes them and
tends to lift the iron core in its interior.

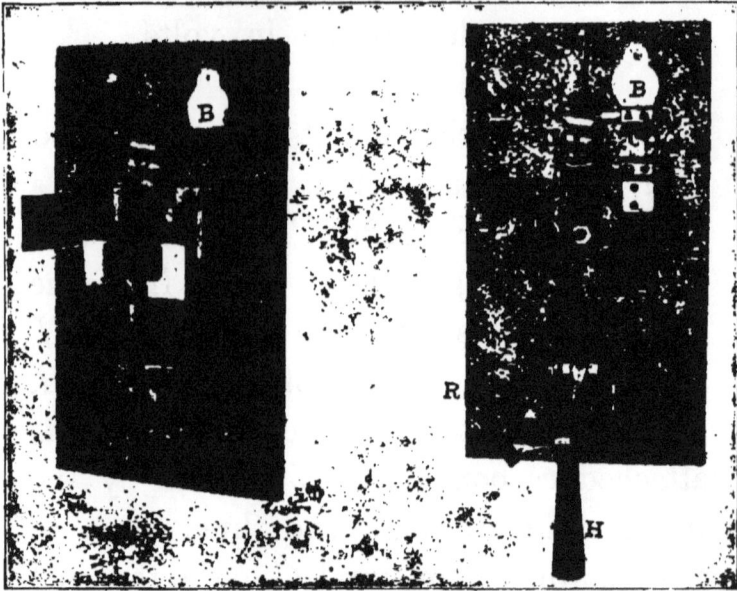

FIG. 126.—CARBON-PLATE AUTOMATIC CIRCUIT-BREAKER.

As soon as the current strength exceeds a
certain limiting safe value, the raising of
the iron core by the increased magnetic

attraction lifts the detent, and permits the switch *H*, to be thrown out of the clips into the position shown on the right hand side. As soon as connection at the clips *P*, *P'*, is broken, a powerful arc would probably form which might melt the switch. Contact is, however, maintained through the medium of the carbon plates *N*, *N*, and the carbon rods *R*, *R*, which brush against them. The arc which takes place when this latter contact is broken is a carbon arc, instead of a copper arc, and such burning as does occur can only result in burning some of the carbon parts, which can be readily replaced from time to time.

Another form of automatic circuit breaker is shown in Fig. 127. Here the circuit is normally closed from the terminal *R*, through the three turns of the spiral *C*, the metallic projections *B*, *B*, and the bridge of flexible copper strips t, t, between them. As soon as the current strength passing through the apparatus exceeds the

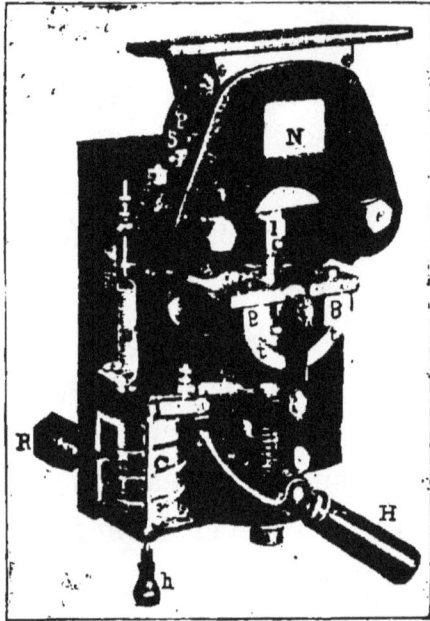

FIG. 127.—MAGNETIC CIRCUIT-BREAKER.

limiting amount for which it is set, the coil C, attracts its armature against the tension of the spiral spring t, and permits

the larger spring S, to withdraw the bridge
t, t, from the blocks B, B. A shunt circuit, is, however, retained between B, B,
for a little while after this contact is
broken through the two magnet coils
M, M, and a smaller set of contacts in the
upper part of the apparatus. The magnets
become powerfully excited by the passage
of the current through them and produce
magnetic poles over the iron surfaces
P, P, P, and P', one pole being, say north,
and the other south. Between these pole
pieces, the second or auxiliary contact is
broken by the descent of the lever l, after
the main contact is broken at B B, and t.
The arc, which tends to follow the inter-
ruption of the auxiliary contact, is instantly
extinguished by the influence of the magnetic flux between the polar projections,
as already explained in the chapter on
controllers.

Should one of the generators, or one of the feeders, become overloaded, the automatic circuit-breaker will open its circuit and protect the generator placed therein. In many cases the overload may have been due to an accidental temporary short-circuit, which almost immediately disappears. In such cases it is usual to reset the circuit-breaker by the use of the handle H, until it is found that after three trials the apparatus refuses to remain set. It is then usual to allow the circuit to remain broken and to search for the short-circuit.

Fig. 128, shows a form of ammeter, such as is seen at M, M, M, in Fig. 125. Here the metallic pieces A, B, form the terminals of the massive coil C, having two turns placed directly in the circuit. The iron core O, is attracted towards this helix, by the electromagnetic action of the current, this attraction increasing with the current strength. The core O, is suspended from a short balance arm pivoted at v, and

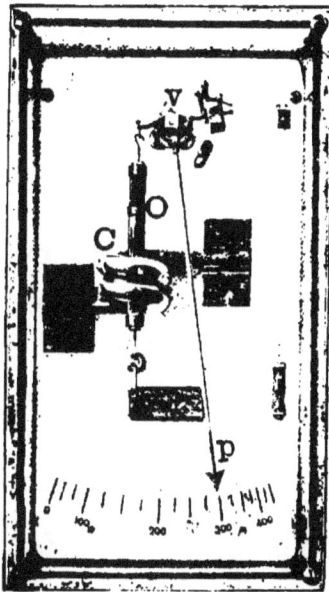

FIG. 128.—FORM OF AMMETER.

having a long pointer or index p, moving over a scale. When the current is cut off, the counterpoise overweights the iron core, and the pointer moves into a position opposite to the zero point on the left hand of the scale. As the current strength through the coil C, increases, the magnetic pull tends to overcome the gravitational pull on the counterpoise, and the pointer moves further and further towards the right.

A form of voltmeter, shown at V, in Fig. 125, is represented on an enlarged scale in Fig. 129. The principle and action of the apparatus are similar to that of the amme-ter in the preceding figure. The principal difference, however, is in the winding of the coil C, which, instead of consisting of but two turns carrying a powerful current, has very many turns carrying a feeble cur-rent. Resistances of insulated wire wound on frames $R\ R$, are placed in circuit with the vertical coil C, and the terminals of the generator. The current strength passing in this circuit will be determined by Ohm's law. For example, if the total resistance of the coil C, and the two resistances R, R,

FIG. 129.—VOLTMETER.

is 5,500 ohms, and the pressure at the gen: erator terminals is 550 volts, then the current strength passing through the circuit will be $\dfrac{550}{5,500}$ volts $= \dfrac{1}{10}$th ampere $= 100$ *milliamperes.* The counterpoise t, is so arranged that at this particular current the pointer p, stands vertical and indicates 550 volts. Should the pressure rise 10 per cent., or to 605 volts, the current in the circuit of the coil C, would increase 10 per cent., and its increased magnetic attraction on the iron core within it would deflect the pointer to a position which is marked 605 volts on the scale. It is evident, therefore, that this voltmeter is essentially an ammeter with a high resistance in its circuit.

The general connection which is effected by the switches on the switchboard, omitting all details of ammeters, voltmeters, cut-outs and lightning arrestors, is diagrammatically represented in Fig. 130.   Here

two *main bars*, or *bus-bars*, *B B*, *B′B′*—a
contraction for *omnibus bars*, so called be-

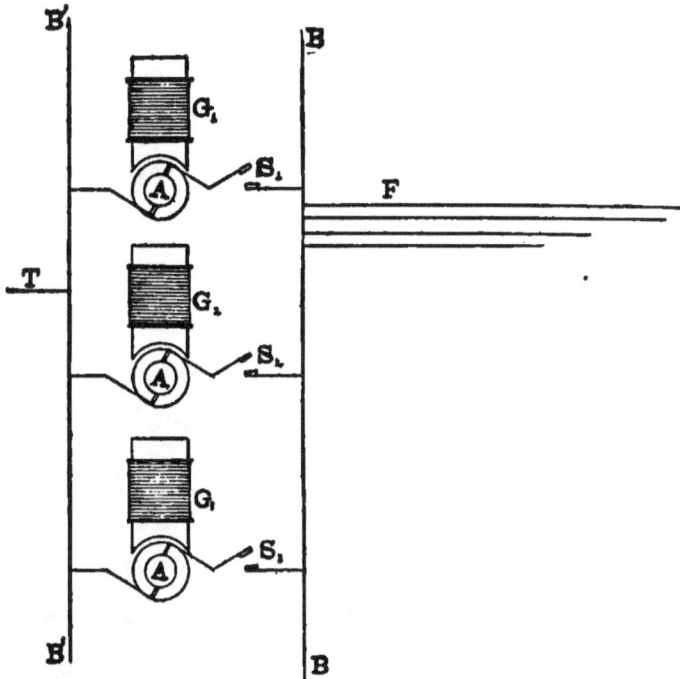

FIG. 130.—GENERAL CONNECTION BETWEEN GENERATORS
AND FEEDERS AT POWER HOUSE.

cause they receive the entire current from
the generators,—are connected, one to the
feeders and the other to the track, ground

feeders, or ground connection. Between these bus-bars the station pressure of say 550 volts is maintained. One or more of the generators G_1, G_2, G_3, are connected across the bus-bars according to the amount of *load* on the lines; *i. e.*, according to the number of cars that are running, and the work they are doing. If only a few cars are on the line the current required will be small, the electric activity small, and a single generator may be sufficient. Thus the switch S_1, may be closed, leaving G_1, to take the entire load. If more cars are run the total current strength supplied to the feeders may require the addition of a second generator G_2, by bringing it up to speed and excitation and closing the switch S_2, and so on for the other generators.

CHAPTER XIII.

GENERATORS AND POWER HOUSES.

TURNING now from the switchboards to the generators which supply them, we notice two distinct types; namely, the *belt-driven generator*, and the *direct-driven generator; i. e.*, a generator directly coupled to the driving engine. The modern tendency in large power houses is to employ very large generators, of say 1,000 HP each, and to connect these directly to a driving-engine. In some power houses, however, belt-driven generators are employed. The belt-driven generators have usually four poles, and very rarely have less than this number. The large direct-driven generators have usually more than four poles, since it is found more convenient and economical to construct generators of large output with a greater number of poles. Fig. 131 shows an example of a belt-driven generator of 500 KW output. Fig. 132 shows a direct-driven generator.

Turning to Fig. 131, N, S, N, S, are the four magnet poles wound with coils of insulated wire. In nearly all cases railway generators are *compound-wound*; *i. e.,* there are two windings on each coil, one of very stout conductor and of very few turns, connected directly in the armature circuit, the other of many turns of fine wire, connected in a shunt, or by-path around the armature. The object of compound winding is to maintain the pressure automatically constant at the brushes, or

FIG. 131.—QUADRIPOLAR BELT-DRIVEN GENERATOR.

at the switchboard bus-bars, notwithstanding changes in the number of cars, or load.

The armature A, revolves within the annular space provided between the four pole-pieces, and with it the commutator C. On the surface of this commutator four sets of *collecting brushes* H, H, are fixed on a frame, capable of slight adjustment in angular position by means of the wheel shown at the base of the pedestal. T, is one of the main terminals, with which the brushes are connected. B, is the driving belt.

In Fig. 132, similar letters refer to similar parts. Here there are also four poles and four sets of brushes, capable of being rotated together within certain limits by the projecting handle. The engine E, is coupled directly to the arma-

FIG. 132.—QUADRIPOLAR DIRECT-DRIVEN GENERATOR.

ture shaft through powerful springs contained within the coupling K. F, is a fly-wheel and P, a cluster of six incandescent lamps in series, called *pilot lamps*.

FIG. 133.—ARMATURE OF DIRECT-DRIVEN GENERATOR.

A particular armature intended for a direct-driven railroad generator is shown in Fig. 133. Here the armature consists of two distinct parts ; namely, a body or core of iron, and conducting wires. The core is *laminated*, that is, formed of a number of thin, soft, sheet-iron discs, provided with slots in their external edges, so that when assembled in the shape of a short cylinder, a number of longitudinal slots or grooves are provided for the reception of the wires. Without considering the winding in detail, it will suffice to say that the conductors are laid in the slots S, S, and are then connected to the separate bars or segments of the *commutator C, C, C*. Fig. 134, shows the operation of winding another form of railway generator armature, with the wires W, W, passing through the slots of the iron armature core A. In this case the commutator is not yet placed on the shaft. A completed armature is, however, shown below at B, with its commutator at C.

During the revolution of the armature through the magnetic flux produced by the field magnets of the generator, E. M. Fs. are induced in the winding, and when their circuit is closed through the feeders produce currents in them. The value of the E. M. F. developed by the armature during its rotation, depends upon the total amount of magnetic flux passing through the armature and its wires, the total number of wires wound over the surface of the armature in the various grooves, and the number of revolutions which the armature makes per minute; *i. e.*, its rotary speed. The current strength which a given armature can maintain steadily, depends upon the size of the wires; *i. e.*, upon the resistance of the armature and its capability of readily disengaging the heat developed by the current in that resistance. The limiting

Fig. 134.—Winding of Multipolar Railway Generator Armature.

current strength is usually determined in practice by the heating of the armature, which in good practice does not exceed 40° C. above the surrounding air, during continuous running.

Illustrations of generator rooms in power-houses, employing respectively the belt-driven and direct-driven types, are shown in Figs. 135 and 136. Fig. 135 shows the interior of the Fifty-second Street power house of the Brooklyn street railway system, containing twelve belt-driven generators, each of 500 KW capacity, capable of a total output of 6,000 KW, and representing 12,000 amperes at 500 volts; or, approximately, 11,000 amperes at 550 volts. These generators are, however, capable of standing a considerable overload for a limited time. The switchboard S, is seen on a gallery at the end of the room.

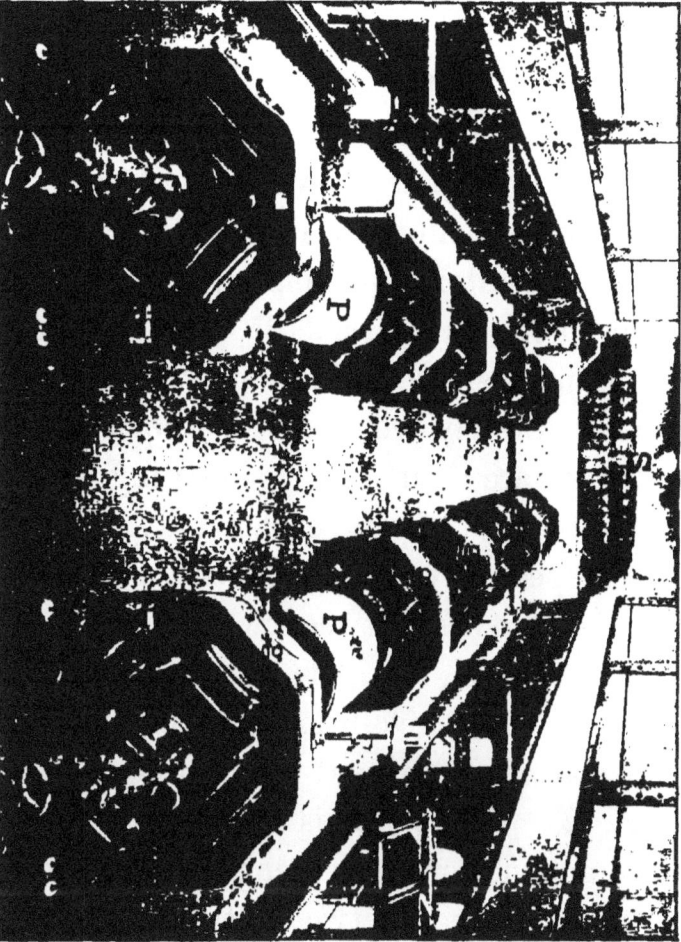

Fig. 135.—500-KW Belt-Driven Generators, Fifty-second Street Power Station, Brooklyn, N. Y.

A view of the interior of another Brooklyn railway power house; namely, that at Kent Avenue, is shown in Fig. 136. Here, instead of being placed on the floor beneath the generators and connected to the latter by belts, as in Fig. 135, the engines are mounted side by side with the generators and directly coupled to the armatures. There are four large generating units in the room of the type shown at 1, 2, 3 and 4 respectively. Each generator has twelve magnet poles, between which revolves the armature A, with its commutator C, at a speed of 75 revolutions per minute. The armature is driven by a double engine E, E. The engine is a double, horizontal, compound-condensing engine, the generator being placed between the two halves. F, F, is the engine fly-wheel placed on the main shaft, close to the armature. Each of these large generators has a

Fig. 136.—1,500-KW Generators, Kent Avenue Power Station, Brooklyn, N. Y.

capacity of 3,000 amperes at a pressure of
550 volts, representing an activity at full
load of 1,650 KW, and a total output,
when all are at full load, of 6,600 KW, or
8,800 HP, approximately. The switchboard *S*, is seen at the end of the room
through the fly-wheels of the two engines,
on the left hand side of the figure. One
of the generators in the figure; namely,
No. 3, is shown incomplete, the field magnets being not yet assembled. The most
recent development in street railway
practice is in the direction of powerful
slow-speed engines and direct-connected
generators of this type.

Fig. 137, shows a plan view of the
engine and generator room in the Delaware
Avenue railway power house at Philadelphia. Here the general plan of engines
and generators is similar to that shown in

FIG. 137.—PLAN OF ENGINE ROOM, DELAWARE AVENUE POWER STATION, PHILADELPHIA.

Fig. 136. There are four generating units marked 1, 2, 3, 4, each consisting of a 1,650 KW, 12-pole generator, with its armature keyed to the shaft of a large compound-condensing engine of 2,000 HP (about 1,500 KW), arranged in two parts, one part on each side of the generator. *S, S,* is the switchboard, behind which the feeders are seen. *P,* is the air-pump connected with all the engines, and 5, is a small 300-KW direct-driven unit for light loads. One of the larger generator units is said to have operated as many as 212 cars at one time. If working at full load this represents a mean activity of 8 KW per car. At this rate all four units could operate 850 cars. Each generator will stand the application of full load without any change in the position of its brushes, and will stand an overload of 50 per cent. with a slight movement of the same.

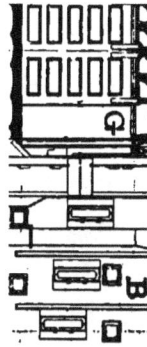

Fig. 138 shows a section of the power house represented in plan in Fig. 137. Here E, E, shows one of the engines, and g, g, one of the large generators on the lower floor. On the floor above are placed the boilers in two rows, one on each side, with an auxiliary G, G, between their fronts. The boiler accommodation is for ten batteries, each of 500 HP, representing an aggregate capacity of 5,000 HP nominally. B, B, are the boilers, shown in cross-section on the right hand and inside view on the left. The steam pipes descend to the ceiling of the generator room, and the engine exhausts are led to the cellar where they are dripped, and the main exhaust pipe X, X, is led to the roof.

CHAPTER XIV.

OPERATION AND MAINTENANCE.

THE amount of power which a street car requires, depends, as we have seen, upon its size, weight, the number of passengers it is carrying, its speed, and the gradient on which it runs. It may vary from no power, when running down hill, to 100 KW when climbing a steep hill. It is often a matter of surprise to those who have been accustomed to see a pair of horses pull a street car through the city streets, that power, representing say more than 100 horses acting together, may be needed on occasions to propel electric cars. The reasons, however, are very clear.   An electric car weighs from 15,000 to 20,000 pounds without passengers, while a horse car weighs only about 5,000 pounds without passengers. The electric car will carry many more passengers than a street car, runs at a greater speed, and will climb grades impossible to be sur-. mounted by two horses.

A good rule to remember is that on the average, over a city street railroad system, an electric street car takes 1 KW for every mile per hour it averages, that is to say, if a car runs at 8 miles per hour it absorbs roughly 8 KW of electric power; or, in 1 hour, would absorb a total amount of work equal to 8 kilowatt-hours. This rough estimate is, of course, independent of the power required to heat the car when electric heating is employed.

The *output* of a station, that is, the load on the generators, varies markedly at different hours of the day. As a rule the heaviest load occurs in the morning and evening hours. The reason for the increased load is not only because a greater number of cars are running and the cars are more heavily laden, but the startings, which require considerable power, occur more frequently during the time of greatest load. Fig. 139 shows *load diagrams* taken in Boston, Mass., on June 16–19, 1895. It will be observed that the load varies from practically 0 at 4 A. M., to 12,000 amperes and 800 cars, and that the total activity correspondingly varies from nearly 0 to about 6,000 kilowatts.

It has been found from a report in 1894, of 232 American electric street railways, operating 5,120 miles of track with a total

capital of \$316,700,000, and a funded debt
of \$279,000,000, that the operating ex-

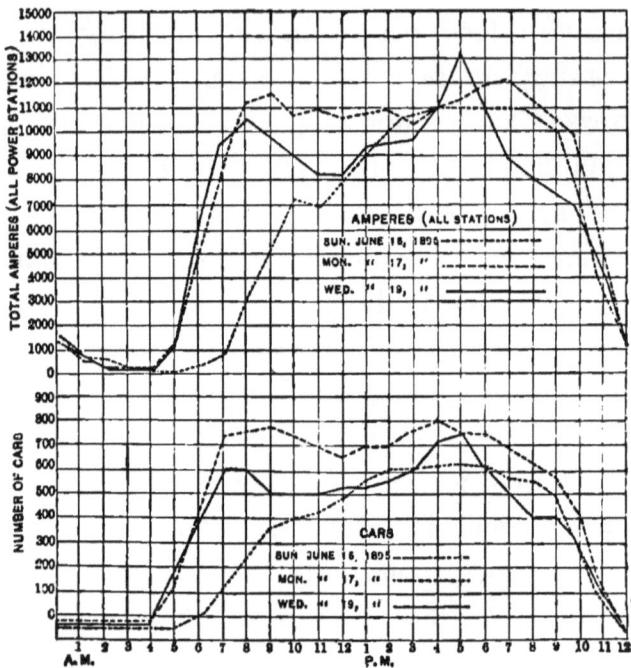

FIG. 139.—LOAD DIAGRAMS AT BOSTON, JUNE 16–19, 1895.

penses were 62.8 per cent. of the gross
receipts, and the fixed charges 22.9 per

cent., leaving a net income of 14.3 per cent. of the gross receipts.

The power required to be installed at the power house varies with a number of local conditions, but averages 20 KW per car in use. The cost of installing this power is about $70 per KW for steam plants, including engines and boilers, and about $30 per KW of combined steam and electric plant, or $100 per KW of total machinery. The cost of the electric equipment of a car, including two 25 HP motors and controllers, is about $1,000, and the cost of a car so equipped complete, roughly $2,300. The line construction costs roughly $5,000 per mile of double track, excluding track construction, but varies considerably, under different conditions. The total expense of a *car mile*, *i. e.*, a run of one mile per car, varies of course considerably with the size, the kind of system and the nature of the traffic, but a fair average may be considered as being from 15 to 25 cents per car mile. Of this the cost of supplying electric power is usually only from 1 cent to 2 cents per car mile. In small systems all these costs are likely to exceed those given.

The size and style of the car which is adopted, varies with the nature of the traffic, and the speed at which the car is expected to run. Under some conditions heavy cars running slowly are desirable, while in others light, high-speed cars are preferable.

The population per mile of street railway track in the United States is, approximately, 4,600, varying between 3,000 in the New England States, and 10,000 in the

Southern States. In Canada, it is about 11,600. The total street car mileage in the United States is about 6 per cent. of the total steam railroad mileage, and the gross earnings about 50 per cent. of the total passenger steam railroad earnings.

For the purpose of facilitating repairs on the line, special wagons drawn by horses are employed, called *tower wagons*, arranged so as to bring the workmen within easy access of the trolley wire. These wagons carry a light platform, which is either rigid or is capable of being raised and lowered. Both the frame of this wagon and its tower being of wood, the men working upon it are practically insulated, except in wet weather.

In latitudes where snow falls the track is kept clear by an electric *snow sweeper*.

One of these snow sweepers is shown in Fig. 140, where the track has been swept by the rotation of the brushes of the car. Fig. 141 shows one of these cars in action.

FIG. 140.—ELECTRIC SNOW SWEEPER.

There are four motors on one of these cars, usually of 25 HP each. Two of the motors are connected with the driving axles in the usual way, and the other two are wound for a higher speed and are connected so as to drive the revolving

FIG. 141.—SNOW SWEEPER IN ACTION.

brushes. These sweeping brushes are fixed at an angle of 45° with the front of the car.

The overhead trolley system has been objected to in cities on account of its unsightliness. The use of trolley poles

with their span, guard, and trolley wires
are certainly far from being a pleasing
ornament to the streets of a well built city.
For this reason attempts have been made
to replace the *overhead trolley system* by
an *underground* or *conduit system* of trolleys, and also by *storage-battery propulsion*.
The overhead trolley system is, however,
considerably more economical to erect and
maintain than either a storage battery or
conduit system.   In large cities, where an
increased cost is preferred to the unsightliness of the overhead trolley system, the
underground trolley may find a successful
use.   It is already being tried in Washing-
ton, D. C., and elsewhere in the United
States, while in the city of Buda Pesth,
Austria, an extended system of underground trolley· roads has been running
successfully for several years.

CHAPTER XV.

STORAGE BATTERY SYSTEMS.

THE admitted unsightliness of the overhead trolley system and the difficulty of
maintaining efficient operation of the
underground trolley, under all conditions
of climate, have led to many efforts to
obtain a *self-contained system* of electric
railways; that is, a system in which each
of the cars will carry its own electric
driving power. In the early history of
the art this was attempted by means of
the primary battery. Primary batteries
are now recognized as being altogether too
expensive for this purpose, owing to the
fact that they derive their motive power

307

from the consumption of zinc in a solution,
a fact which will effectually prevent such
batteries from competing with other types
of motive power so long as the price of
the zinc, and the solution in which it is
dissolved, maintain anything like their
present values.

The nearest approach to the successful
solution of the problem of an electrically
propelled car, which carries its own stored
electric energy, is found in the use of the
secondary or storage cell. In this system
the storage cells derive their charge, or
stored electric energy, from electric current supplied to the cells at some central
station. As some time is required to
charge the cells, they are usually removed
from the car to receive their charge.
Before proceeding to the general description of the storage battery equipment of a

car, a brief account of the construction and operation of storage batteries will be necessary.

A great variety of forms have been given to the *secondary or storage cell*. In practically all cases, the material of which their *plates* or *elements* are formed is lead. If two sheets of lead be immersed in a solution of dilute sulphuric acid, and an electric current be sent through the solution from one plate to the other, an electrolytic decomposition will occur, whereby the *positive plate*, or the plate at which the current enters, becomes oxidized, while the *negative plate*, or that at which the current leaves the cell, liberates bubbles of hydrogen gas. During this process a C. E. M. F. is set up in the cell amounting, probably, to about 2.5 volts, and every coulomb, or ampere-second of electricity, which passes through the cell, does work in it amounting to 2.5 volt-coulombs or 2.5 joules. At a rate of 1 ampere, or 1 coulomb per second, the work so expended in the cell would amount in one hour, to $3,600 \times 2.5 = 9,000$ joules or 2.5 *watt-hours*.

If there were no resistance in the cell; and if, moreover, no free hydrogen gas escaped from it, all the above work would be expended in chemical action, which would be stored up in the cell in the form of chemical products. So far as the C. E. M. F. is due to the drop of pressure through the resistance, the work is expended as heat, but so far as it is produced by the C. E. M. F. of chemical action, it is theoretically possible to store the work in chemical combinations. If after having been charged in this way the cell is removed from the charging circuit and its plates are connected through a wire, it will act as a primary battery; that is to say the oxidized plate will behave like the copper plate of an ordinary bluestone cell, and the unoxidized plate like the zinc of such a cell.

During discharge, the E. M. F. of the cell may, perhaps, average 2 volts, and each coulomb of electricity supplied through the circuit by this E. M. F. represents a delivery of 2 joules of work. During discharge, and the performance of work, the surface of the oxide on the positive plate becomes partially deoxidized, while the plain lead or negative plate becomes partially oxidized. Finally, when the cell is completely discharged, the two plates are superficially the same, each being partially oxidized. A cell is, however, never permitted to completely discharge. In order to restore the cell to its active condition, it is necessary to once more charge it by passing through it the requisite quantity of electricity.

In the case of a primary cell, in which the two plates or elements have essentially different chemical composition, the complete discharge is accompanied by the consumption of one of the plates; namely, the zinc plate. It is impossible, in practice, to restore the active condition of the primary cell by sending a charging current through it, and the plates have to be renewed. In the secondary cell, instead of renewing the discharged plates, the electric current is permitted to reverse the chemical changes which have accompanied discharge and thus restore the active condition.

Instead of using plain lead plates, special forms of lead plates are employed to expose a very large surface to the active liquid. A form of storage cell is

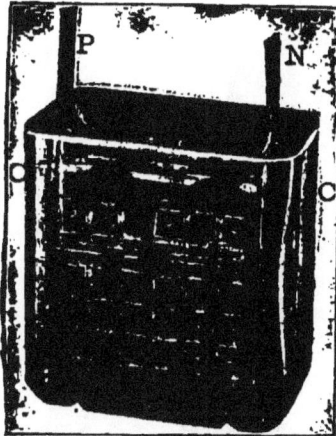

FIG. 142.—FORM OF STORAGE CELL.

shown in Fig. 142. Here the glass cell or jar *C, C*, contains seven flat plates, three of which are connected with the positive terminal *P*, and four to the negative terminal *N*. The solution of sulphuric acid and water is poured in until the plates are covered.

A positive plate is shown in Fig. 143. Here thirty-nine circular buttons, or discs

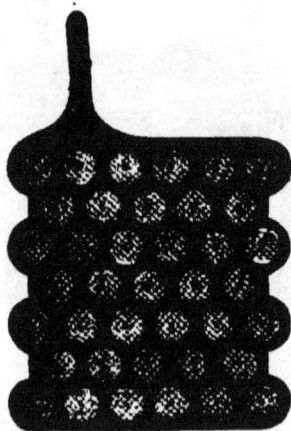

FIG. 143.—POSITIVE PLATE.

of peroxide of lead, are held tightly in a *frame* or *grid* of antimonous lead. The addition of antimony in sufficient quantity prevents the lead grid from being chemically attacked by the solution during charge or discharge. Fig. 144 shows a negative plate, with sixty-four square buttons of soft porous or spongy lead

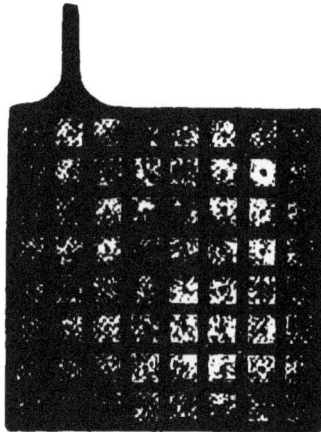

FIG. 144.—NEGATIVE PLATE.

similarly held in an antimonous lead frame. The small holes in the centres of the buttons play no part in the action of the cell, and are made during the mechanical construction of the buttons.

The principal difficulty which has been encountered with the use of storage cells in electric traction, has been in the electric overloads which have sometimes been necessary, and which greatly decrease the life of the plates. If the cars invariably ran upon a level grade and their load remained uniform, it would not be a difficult matter to ensure an absence of electric overloads, or undue calls for power upon the batteries. In practice, however, owing to the existence of curves and grades and over-discharging, the cells are generally soon injured, so that their maintenance becomes very expensive. Moreover, the great weight of the batteries adds largely to the non-paying weight of the car. Considerable improvements have, however, recently been effected in the storage battery whereby better results may be expected.

A form of *storage battery car truck* at
present in use on Madison Avenue, New
York City, is shown in Fig. 145. Here
by turning the motors outwards towards
the ends, that is supporting them on the
opposite side of the axle to that usually
adopted, the space *A B C D*, is reserved

FIG. 145.—STORAGE BATTERY TRUCK.

in the centre of the truck for the reception of the storage battery. A truck with
a storage battery in place is shown in Fig.
146. In this truck sixty storage cells are
arranged in two batteries of thirty cells
each. Since the mean E. M. F. of discharge in a storage cell is, approximately,

2 volts, this represents a pair of batteries each having an E. M. F. of 60 volts. Each cell has 400 ampere-hours capacity; that is, is capable of supplying 40 amperes for 10 hours, or 20 amperes for 20 hours, or 10 amperes for 40 hours, etc., the total quantity of electricity being $400 \times 3,600 =$

FIG. 146.—CAR TRUCK WITH BATTERIES IN PLACE.

1,440,000 coulombs. The above mentioned 1,440,000 coulombs, representing as they do the capacity of its battery, should, theoretically, be discharged whether the duration of discharge is long or short, that is to say, whether the cells are allowed to discharge in a few minutes or in many hours.

In practice, however, there is always a marked diminution in the available quantity of electric discharge when the duration is too brief, say below three hours. If the E. M. F. of discharge averages 2 volts, the total amount of energy available from each cell is 2 × 1,440,000 = 2,880,000 coulomb-volts, or joules, and 60 such cells should hold a total quantity of energy of 172,800,000 joules. Since 1 watt-hour is 3,600 joules, and 1 KW hour 3,600,000 joules, the total energy in the battery is 48 KW-hours. Consequently, the activity of the battery, assuming no loss, by very rapid discharging, would be 8 KW maintained for six hours, or 12 KW maintained for four hours. Of this power some will necessarily be lost in the motors and gears, so that, perhaps, only about 75 per cent. may be available at the car axles.

Fig. 147 shows diagrammatically the connections obtained in the different positions of the controller of this car. In position 1, the two batteries are placed in parallel, making an effective E. M. F. of 60

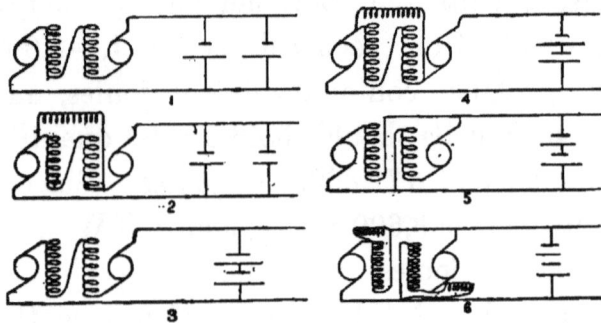

FIG. 147.—CONTROLLER POSITIONS.

volts at main terminals, while the two motors are in series, each motor receiving 30 volts. If under these conditions, the activity of the battery is 12 KW, the current strength received by the two motors in series will be $\dfrac{12,000}{60} = 200$ amperes.

In the second position, a shunt is thrown around the field magnets of the motors, thereby diminishing their magnetic power, and requiring a greater speed from the armatures in order to develop the necessary C. E. M. F. of 60 volts in all.

In the third position, the two batteries are thrown in series, representing a total E. M. F. available at terminals of 120 volts, and a corresponding increase in the speed of the unshunted motors to produce this C. E. M. F.

In the fourth position, a shunt is again thrown around the field magnets of the two motors, and their speed is correspondingly increased.

In the fifth position, the two unshunted motors are thrown in parallel, instead of in series, thus calling upon each motor to de-velop a total C. E. M. F. of 120 volts.

In the sixth and last position, the mag-nets of the motors are shunted, requiring the armatures to run faster in order to pro-duce 120 volts total C. E. M. F. in the motor under these conditions.

When the car returns to the car house and the battery has been sufficiently dis-charged, it is lifted bodily from the truck and replaced by a charged battery.

CHAPTER XVI.

ELECTRIC LOCOMOTIVES.

WITHIN large cities, municipal ordinances generally limit the speed of street cars to about eight miles per hour. In suburban districts, however, a speed is usually permitted as high as fifteen miles per hour, while in inter-urban traffic, speeds of thirty miles per hour or more are sometimes reached. As the velocity of the cars increase, the electric activity which must be supplied to them increases in nearly the same proportion ; for, the torque exerted by the motors on a given gradient remains nearly the same at all the above mentioned speeds, the rate only varying at which that torque is exerted.

At still higher speeds than the preceding, the friction between axles and journals, and the wheels and the track, does not sensibly increase, but the friction between the surface of the car and the air does sensibly increase, so that, at speeds above 100 miles per hour, the track and journal friction would probably commence to be small compared with the resistance to air displacement and friction. Consequently, for very high speeds, the form of the moving car becomes nearly as important as the form of the hull of a steamer; only in the case of the latter, the hull only is exposed to the friction against the water, while in the case of the car, the entire surface is moved through the air.

The question has often arisen as to the early probability of replacing steam propulsion on ordinary railroads by electric propulsion. The schedule speeds of ex-press trains on steam roads have altered but little during the last twenty years, judging from an inspection of railroad time tables included in that period. There is no doubt, however, that the introduction of the electric locomotive would permit much higher speeds to be safely attained, and, when this fact is taken in connection with the manifest advantages possessed by electric propulsion, it would seem that in electricity, steam has a formidable rival in this field. The question, however, is one of public demand, and economy of transportation. There can be no doubt, that so far as regards economy in long-distance transportation, steam propulsion is cheaper than electric propulsion, owing to the cost of the plant, since the cost of transmitting power electrically increases rapidly with the distance. Consequently, for freight and slow

traffic, it does not seem that the immediate
future will witness the displacement of the
steam locomotive, but for high-speed passenger transportation, the extra cost of the
electric equipment may be repaid by the
increased economy in time of transit, so
that it does not seem improbable that in
the near future the high-speed passenger
locomotive may come into use on railroads.

As an example of experiments which
have been tried in the direction of high-
speed electric railroads, we may mention
the bicycle railroad shown in Fig. 148.
Here the car runs on a single rail and rests
on two wheels, which, instead of being
placed side by side, as in the ordinary
truck, are in the same plane, like a bicycle,
one being placed in the front and the other
in the rear. The ends of the car are tap-
ered, as shown. To prevent the car from

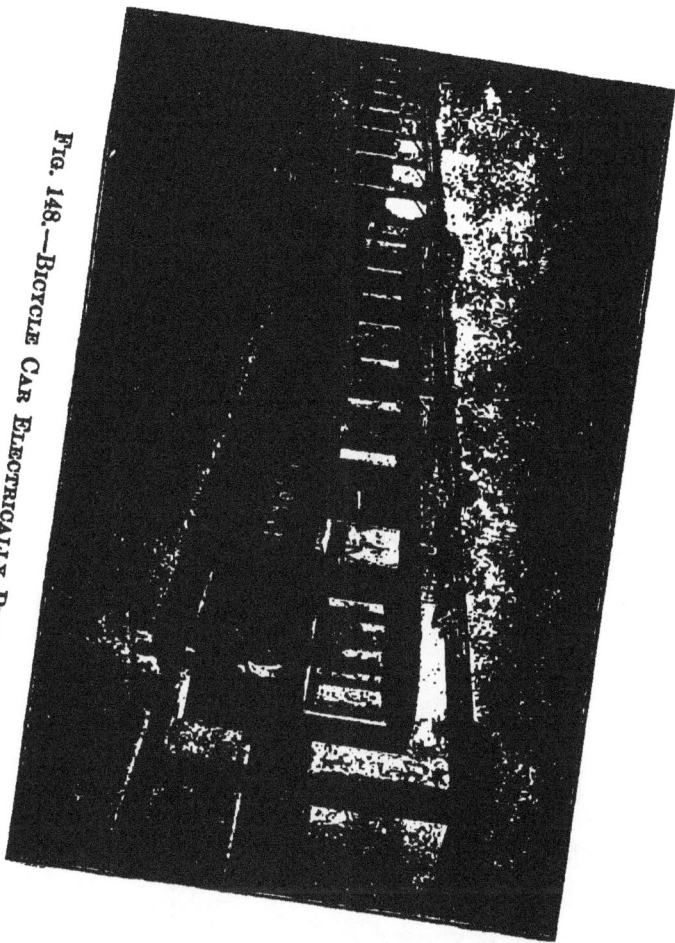

FIG. 148.—BICYCLE CAR ELECTRICALLY PROPELLED.

falling sideways when at rest, it is supported by guide wheels pressing upon the upper or guide rail, which serves the double

FIG. 149.—SECTION OF BICYCLE CAR.

purpose of a support and an electric conductor.   A cross-section of a double deck car is shown in Fig. 149.   It will be seen that these cars are only of half width,

two being able to pass each other with nine
inches clearance within the space occupied
by an ordinary 4′ 8 1/2″ track. The advantage claimed for this construction is
that it not only enables the traffic to be
doubled upon any existing railroad by
erecting the upper or trolley guides, one
for each existing rail, but it also enables
the weight of the cars to be materially
reduced, since the narrow car enables
the necessary structural strength to be
obtained with less material, and the
weight of the loaded car, per passenger
carried, would be about four times less than
with the existing construction, thus econo-
mizing in activity expended against journal
friction and grades. The electric propulsion is obtained from a single motor M, in
the front wheel of the car. On the track
shown in the figure, speeds of 45 miles per
hour are readily obtained, and speeds of

over 60 miles an hour are claimed to have been reached on a track 1 1/2 miles in length. By giving a lean to the upper or guide rail no difficulty has been found in going around sharp curves, since no appreciable strain is produced. A disadvantage of the system is that it can only provide seats for two in the width of the car.

Another purpose to which the electric locomotive has already been applied is to the drawing of trains of cars through long tunnels on steam roads. As is well known considerable difficulty is experienced in ventilating long tunnels when steam locomotives pass through them frequently. This difficulty is entirely overcome by the use of the electric locomotive. Here the requirements are not for high speed, but for a powerful draw-bar pull. An example of this type of electric locomotive is seen in the Belt Line Tunnel at Baltimore. This tunnel is about a mile and a half long, and has a gradient of about forty-two feet to the mile. Since the freight traffic is heavy, a powerful locomotive is required to draw the trains. Fig. 150 shows the entrance to the tunnel with the electric overhead conductors C, C, in place. One of these conductors is provided for each of the two tracks. w, w, are the copper supply wires, and K, K, are the supporting catenaries or rod chains.

Fig. 151 shows one of the conductor supports from the catenary. r, r, are the rods of the catenary. I, is the conical insulator. R, R, the suspension rods from this insulator. B, B, is the beam supported by these rods, and C, C, the conductors which are formed of iron bars, arranged opposite

Fig. 150.—The Entrance to the Tunnel.

each other, so as to leave a slot between them and enclose an inverted conduit. In this conduit slides a brass shoe supported

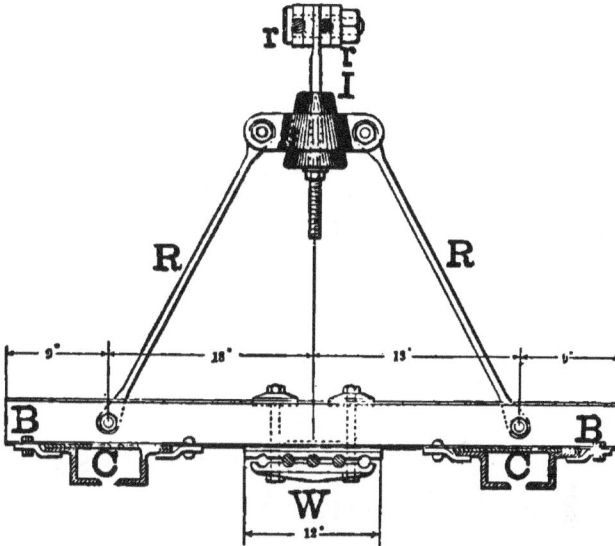

FIG. 151.—OVERHEAD CONDUCTOR SUPPORT.

on a flexible rod from the top of the loco-motive. *W*, is a cross-section of the sup-ply wires or feeders, which are stranded

copper cables about one inch in diameter
clamped directly to the beam as shown.

The method of supporting the conductors in the tunnel is shown in Fig. 152.
Here *M, M, M,* is the masonry arch of
the top of the tunnel, *B, B,* are bolts let

FIG. 152.—METHOD OF SUPPORTING CONDUCTORS IN THE
TUNNEL.

into the masonry, and supporting a chan-
nel frame by two conical insulators *i, i,*
at the ends. Two other insulators *i″, i″,*
support the conductors *c, c.*

Fig. 153 shows the electric locomotive
pulling a steam locomotive and train
through the tunnel. *F, F,* is the flexible

FIG. 153.—THE BALTIMORE AND OHIO RAILROAD COMPANY'S NINETY-SIX-TON ELECTRIC LOCOMOTIVE AND FREIGHT TRAIN LEAVING THE NORTH EXIT OF THE BALTIMORE BELT LINE TUNNEL.

conductor corresponding to the trolley pole
of an ordinary street car, and carrying at
its extremity the shoe running in the conductor overhead. An end view of the
locomotive is shown in Fig. 154. The
trolley fastened to the top of the locomotive is shown in side and end view at Fig.
155. *S*, is the shoe, and *J, J*, the joints in
the structure, which automatically lengthen
and shorten the trolley pole to conform
with the varying height of the trolley conductor. This locomotive weighs ninety-
six short tons in all, and is supported on
two trucks and four pairs of driving
wheels. A motor is mounted directly
on each driving axle, thus placing four
motors in the locomotive. One of
these motors is shown in Fig. 156.
Here the iron-clad armature *A, A,* is
mounted in a sextipolar field frame *F, F.*
These motors being mounted on the driv-

FIG. 154.—END VIEW OF ELECTRIC LOCOMOTIVE.

ing axles through special flexible connections without the intervention of gears, are
called *gearless motors*. The method of
mounting them in the truck is shown in

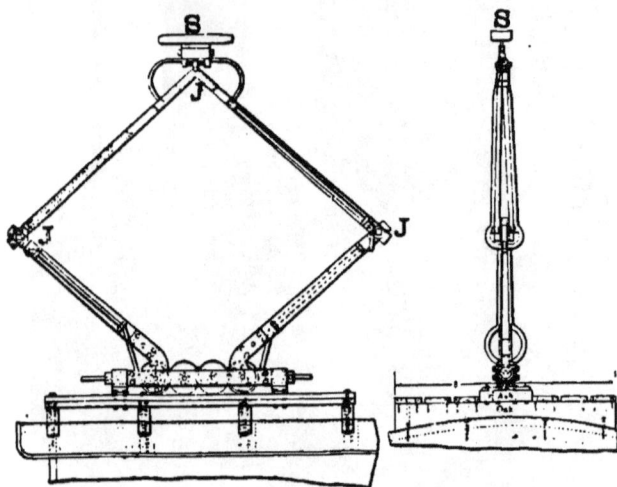

FIG. 155.—SIDE AND END VIEWS OF TROLLEY.

Fig. 157. Here S, S, is the side frame
J, J, are the journal boxes of the two
axles in the truck, and M, M, the motors
mounted flexibly over each axle.

The current is supplied to each motor armature through six pairs of carbon

FIG. 156.—MOTOR OF ELECTRIC LOCOMOTIVE.

brushes arranged around the periphery of the commutator. The total current supplied to each motor is normally about 500 amperes at full load. The pressure of supply is about 600 volts. The normal activity absorbed by each motor at full load is, therefore, 300 KW, or, roughly, about 400 HP. Since there are four motors, this powerful locomotive absorbs

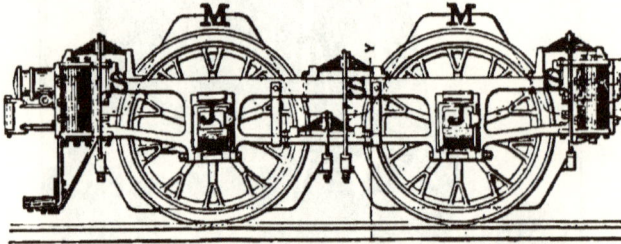

FIG. 157.—TRUCK, SHOWING MOTORS IN POSITION.

a total activity of about 1,600 HP, and the locomotive is rated at 1,500 HP. The locomotive is designed so as to exert a steady pull of 40,000 pounds, or 20 short tons, at the draw bar when drawing a train twelve miles per hour. This represents a useful activity of 1,280 HP in addition to that required to move the locomotive itself. The maximum available draw-bar pull is stated to be 60,000 pounds. The

FIG. 158.—" TERRAPIN BACK " ELECTRIC MINING LOCOMOTIVE.

draw-bar pull in an electric locomotive is uniform, whereas the draw-bar pull in the steam locomotive is necessarily variable at different portions of the stroke. The draw-bar pull of a powerful 60 short-ton steam engine does not usually exceed 25,000 pounds.

Electric traction has recently been adopted on two short branches of road in connection with steam railroads. These are at Nantasket Beach, Mass., and Mount Holly, N. J. The road between Mount Holly and Burlington is about seven miles long, and is operated by electric cars equipped with 100 HP motors; the speed being about thirty miles an hour, and the schedule time for the trip twenty-one minutes, including stops. It is not at all improbable that this is but the beginning of an extensive use of electric traction for suburban traffic, in connection with steam railroads.

The electric locomotive has recently

found a field of application in mining operations. It is especially fitted for such work from the ease with which it is controlled. Fig. 158 shows a form of mining locomotive suitable for hauling trains of trucks through the galleries of a mine. It will be noticed that the trolley pole is of the same general type as that described in connection with the locomotive of the Baltimore tunnel.

THE END.

INDEX.

A

B

C

F

G

M

N

O

Ohm, 35.

——, Practical Definition of, 41.

Ohm's Law, 42.

Open Car Wheels, 107.

—— Circuit, 27.

Operation and Maintenance, 297 to 306.

Output of Station, 299.

P

Page, 9.

Panel, Pressure, 263.

Panels, Feeder, 263.

Parallel Connection of Street Cars, 187, 188.

Permanent Horseshoe Magnet, 83.

Pilot Lamps, 284.

Pinion Armature, 115 to 117.

——, Double, 119.

——, Rawhide, 120.

Pinkus, 9.

Plate of Storage Cell, 309.

Pneumatic Car Brake, 122.

Points, Feeding, 65.

Pole, 32.

—— Climbers, 224.

——, Trolley, 205.

S

T

THIRD EDITION. GREATLY ENLARGED

A DICTIONARY OF

Electrical Words, Terms, and Phrases.

By EDWIN J. HOUSTON, Ph.D. (Princeton).

AUTHOR OF

"Advanced Primers of Electricity"; "Electricity One Hundred Years Ago and To-day," etc., etc.

Cloth, 667 large octavo pages, 582 illustrations, Price, $5.00.

Some idea of the scope of this important work and of the immense amount of labor involved in it, may be formed when it is stated that it contains definitions of about 6000 distinct words, terms, or phrases. The dictionary is not a mere word-book; the words, terms, and phrases are invariably followed by a short, concise definition, giving the sense in which they are correctly employed, and a general statement of the principles of electrical science on which the definition is founded. Each of the great classes or divisions of electrical investigation or utilization comes under careful and exhaustive treatment; and while close attention is given to the more settled and hackneyed phraseology of the older branches of work, the newer words and the novel departments they belong to are not less thoroughly handled. Every source of information has been referred to, and while libraries have been ransacked, the note-book of the laboratory and the catalogue of the wareroom have not been forgotten or neglected. So far has the work been carried in respect to the policy of inclusion that the book has been brought down to date by means of an appendix, in which are placed the very newest words, as well as many whose rareness of use had consigned them to obscurity and oblivion.

Copies of this or any other electrical book published will be sent by mail, POSTAGE PREPAID, *to any address in the world, on receipt of price.*

The W. J. Johnston Company, Publishers,

253 BROADWAY, NEW YORK.

Gerard's Electricity. With chapters by Dr. Louis Duncan, C. P. Steinmetz, A. E. Kennelly and Dr. Cary T. Hutchinson. Translated under the direction of Dr. Louis Duncan..................................... $2.50

The Theory and Calculation of Alternating-Current Phenomena. By Charles Próteus Steinmetz 2.50

Central Station Bookkeeping. With Suggested Forms. By H. A. Foster......................... 2.50

Continuous Current Dynamos and Motors. An Elementary Treatise for Students. By Frank P. Cox, B. S. 271 pages, 83 illustrations............... 2.00

Electricity at the Paris Exposition of 1889. By Carl Hering. 250 pages, 62 illustrations. 2.00

Electric Lighting Specifications for the use of Engineers and Architects. Second edition, entirely rewritten. By E. A. Merrill. 213 pages............. 1.50

The Quadruplex. By Wm. Maver, Jr., and Minor M. Davis. With Chapters on Dynamo-Electric Machines in Relation to the Quadruplex, Telegraph Repeaters, the Wheatstone Automatic Telegraph, etc. 126 pages, 63 illustrations....................................... 1.50

The Elements of Static Electricity, with Full Descriptions of the Holtz and Topler Machines. By Philip Atkinson, Ph.D. Second edition. 228 pages, 64 illustrations.................................... 1.50

Lightning Flashes. A Volume of Short, Bright and Crisp Electrical Stories and Sketches. 160 pages, copiously illustrated............................. 1.50

A Practical Treatise on Lightning Protection. By H. W. Spang. 180 pages, 28 illustrations, 1.50

Experiments With Alternating Currents of High Potential and High Frequency. By Nikola Tesla. 146 pages, 30 illustrations.......... $1.00

Lectures on the Electro-Magnet. Authorized American Edition. By Prof. Silvanus P. Thompson. 287 pages, 75 illustrations.......... .... 1.00

Dynamo and Motor Building for Amateurs. With Working Drawings. By Lieutenant C. D. Parkhurst 1.00

Reference Book of Tables and Formulæ for Electric Street Railway Engineers. By E. A. Merrill................................... 1.00

Practical Information for Telephonists. By T. D. Lockwood. 192 pages..................... 1.00

Wheeler's Chart of Wire Gauges........... 1.00

A Practical Treatise on Lightning Conductors. By H. W. Spang. 48 pages, 10 illustrations. .75

Proceedings of the National Conference of Electricians. 300 pages, 23 illustrations.......... .75

Wired Love ; A Romance of Dots and Dashes. 256 pages.. .75

Tables of Equivalents of Units of Measurement. By Carl Hering........................ .50

Copies of any of the above books or of any other electrical book published, will be sent by mail, POSTAGE PREPAID, *to any address in the world on receipt of price.*

THE W. J. JOHNSTON COMPANY,
253 BROADWAY, NEW YORK.